"Lushenko and Raman make a vital contribution to the debate over public opinion and drone use. Getting us beyond simple questions of approval or disapproval, their research opens important doors to rigorous exploration of how public opinion about legitimacy in the use of military force is shaping and being shaped by drone use, including in cross-national comparative perspective. Their highly original research moves the dial on what we can and should be doing to better understand how drone use is impacting the present and future uses of military force and the role democratic publics play in keeping that within the bounds of legitimacy."

John C. Williams, *Durham University, UK*

"The legitimacy of drone warfare is a highly contentious debate to which Paul Lushenko and Shyam Raman have added a fresh perspective. Their book introduces and tests an elegant theory that tracks how evolving patterns of drone warfare shape public perceptions of legitimacy. An important step in the quest for global governance of drones, *The Legitimacy of Drone Warfare* is a must read for policy makers, academics, and drone critics alike."

Daniel Brunstetter, *University of California, Irvine, USA*

"This is the best book I have reviewed that analyzes public attitudes about drones, which have irrevocably altered the character of war and will continue to have ever-increasing impacts. It is theoretically innovative, offering a new understanding of drone warfare. It is methodologically rigorous, using novel experiments and simulations to show how evolving patterns of drone warfare shape public perceptions of legitimacy around the world. And, it is relevant, bridging the theory and practice of drone warfare to inform policy, strategy, and research. Any serious practitioner or policy-maker needs to read this defining book."

Richard Clarke, *General Retired, former Commander of the U.S. Special Operations Command, USA*

"In a crowded field of studies on drone strikes, this new book adds valuable insight by homing in on an often-neglected concept: perceived legitimacy. It is a must-read for scholars who work on air warfare as well as those who study attitudes toward war."

Janina Dill, *Oxford University, UK*

"This book is a model of research on one of the most consequential developments in international security this century: drone warfare. Lushenko and Raman develop and empirically test an original interdisciplinary theory, informed by unique combat experiences, that shapes how we understand the legitimacy of drone warfare. In doing so, they advance the debate on emerging technologies in war and their findings have important implications for U.S. foreign policy and military strategy."

Barry McCaffrey, *General Retired, U.S. Army, USA*

"The legitimacy of the use of drones in war and counterterrorism has emerged as a key issue in academic research and policy debates over the past two decades. Lushenko and Raman's work brings clarity to the issues at stake by forcing us to think clearly about how legitimacy functions in international security. By contributing new empirical evidence that tests our baseline assumptions regarding the legitimacy of drone use, this book makes a valuable contribution to the study of drone warfare and will be an important future reference point for the field."

Jack McDonald, *Director, Centre for Science and Security Studies, Kings College London, UK*

"Lushenko and Raman provide a novel and extremely pertinent framework for evaluating drone warfare as it relates to public perception and legitimacy. Legitimacy is everything in conflict. Legitimacy matters for the politician that orders an action, the practitioner that executes the action, and the public that forms an opinion of the action. The implications of Lushenko and Raman's framework, while extremely important to drone warfare, has far greater reach that extends into future conflicts as we move into the age where artificial intelligence, autonomy, and robotics will take a larger role in fighting wars. Highly recommended read for anyone involved in international relations, national security, and government policy."

Wayne Phelps, *Lieutenant Colonel Retired, U.S. Marine Corps, USA*

"*The Legitimacy of Drone Warfare: Evaluating Public Perceptions* makes a sophisticated and original contribution to our understanding of the crucial concept of legitimacy in international relations by examining the subtle factors that shape public attitudes toward the use of drone warfare. This analysis is especially valuable because of the controversy surrounding the use of such aircraft and the ways in which public perceptions of legitimacy shape how states deploy them. The result is a set of rich insights not only on drone warfare but on the complex relationship between state military operations and the conditions under which they enjoy public support or opposition."

Mitt Regan, *Georgetown University, USA*

"Lushenko and Raman provide an illuminating study on the public's perceptions of drone warfare. By utilising drones as a lens to analyze contemporary conflict, the authors offer an original take on how impactful drone warfare has been to shaping the public's understandings of legitimacy in the lethal deployment of force. A must read for political officials, military personnel, and all those interested in the growing global use of military drones."

James Rogers, *Executive Director, Tech Policy Institute at Cornell University, USA*

The Legitimacy of Drone Warfare

This book examines public perceptions of the legitimacy of drones and how this affects countries' policies on and the global governance of drone warfare.

Scholars recognize that legitimacy is central to countries' use of drones, and political officials often characterize strikes as legitimate to sustain their use abroad. This book introduces and tests an original middle-range theory that allows scholars, policy-makers, and practitioners to understand how evolving patterns of drone warfare globally shape the public's perceptions of legitimacy that can moderate countries' drone policies and the global governance of drones. Rather than relate drone warfare to a platform or counterterrorism strikes only, as experts often do, this book argues that drone warfare is best understood as a function of the unique ways that countries use and constrain strikes. By updating theories of drone warfare, this book provides a generalizable way to understand public perceptions of legitimacy in cross-national contexts, especially among democratic political regimes that are prefigured on political officials' accountability for the use of force abroad.

This book will be of interest to students of security studies, foreign policy, media and communication studies, and international relations.

Paul Lushenko is Assistant Professor and Director of Special Operations at the U.S. Army War College.

Shyam Raman is Visiting Assistant Professor at Williams College.

The Legitimacy of Drone Warfare

Evaluating Public Perceptions

Paul Lushenko and Shyam Raman

LONDON AND NEW YORK

First published 2024
by Routledge
4 Park Square, Milton Park, Abingdon, Oxon OX14 4RN

and by Routledge
605 Third Avenue, New York, NY 10158

Routledge is an imprint of the Taylor & Francis Group, an informa business

British Library Cataloguing-in-Publication Data
A catalogue record for this book is available from the British Library

ISBN: 978-1-032-61428-1 (hbk)
ISBN: 978-1-032-61427-4 (pbk)
ISBN: 978-1-032-61426-7 (ebk)

DOI: 10.4324/9781032614267

Typeset in Times New Roman
by Apex CoVantage, LLC

Contents

Figures

Tables

Acknowledgments

The genesis of this book is a fortuitous meeting between Shyam and me in the fall of 2021, while both of us completed a course taught by Professor Suzanne Mettler at Cornell University entitled "The Politics of Inequality in the United States." Shyam took the course to broaden his understanding of inequities embedded within the United States' social safety net, which relates to his research on topics at the intersection of health, public, and urban economics. I took the course to fulfill a key requirement for my doctoral degree, though what I learned about intergenerational inequity in the United States further shaped my research on the socio-economic effects of unmanned aerial vehicles, or drones, as they are commonly called. Indeed, our initial conversation in Suzanne's class revealed shared research interests for the diverse implications of emerging technologies on the modern battlefield.

Shyam was interested in exploring the economic impacts of U.S. drone strikes, especially in undeclared theaters of operations such as Pakistan, which are not internationally authorized and do not (ostensibly) have U.S. boots on the ground. Yet he lacked an understanding of these operations as well as the drone warfare scholarship that I could provide. I had previously drafted a design-based identification strategy to study the effectiveness of shifting—reasonable certainty and near-certainty—targeting standards for U.S. drone strikes, informed by my combat experiences managing hundreds of these operations around the world. Despite, or perhaps because of, this innovative approach, I sought Shyam's mentorship to learn additional econometric techniques that could allow me to make novel contributions to the renewed literature on drones. Thus, complementary expertise, curiosity, and desire for professional growth conditioned our ongoing exchanges for the better part of two years and constitute the cornerstone of this project.

Though it is true that the theoretical impetus for this book derived from my dissertation research, itself about how we understand the public's perceptions of legitimate drone warfare in a comparative context, this book has benefitted from years of collaborative research on the evolution of drone warfare globally as well as the implications for public attitudes. The conceptual framework in which we define drone warfare as a function of three observable strike attributes, including countries' varying use and constraint of strikes to help minimize unintended consequences, namely civilian casualties, is related to my interest in further problematizing this emerging form of warfare after nearly two decades of waging it. The empirical approach, especially for Chapters 4 and 5, benefited greatly from Shyam's econometric training. At the same time, both of us have presented this research at conferences, seminars, and lectures around the world, receiving helpful feedback that has enhanced the rigor of our analysis. This, in addition to several anonymous reviewers, has strengthened the quality of our book.

Indeed, we incur an enormous debt of gratitude for writing this book, which we can never possibly repay. Speaking for myself, I am privileged to have an extraordinarily devoted wife and children, whose support was instrumental in the research and analysis for this project. I am also thankful to a superb mentor, Professor Sarah Kreps, whose advice has made a real difference in shaping this book into something meaningful. Finally, the U.S. Army, consisting of the soldiers I have led, the peers I have learned from, and the commanders I have worked for, has also been integral to the completion of this project. Specifically, I want to thank Lieutenant Colonel (retired) Keith Carter, Ph.D., for his abiding friendship. He's been the quintessential "Ranger Buddy," providing counsel on how to reconcile thinking and doing in the U.S. Army. For his part, Shyam wants to thank his supportive mentors—Professors John Cawley and Nicholas Sanders—for playing a large role in shaping the empirical rigor and approach of this work.

The views presented in this volume are the authors' alone. They do not represent the policy or position of the U.S. government, Department of Defense, or Army.

Paul Lushenko
Shyam Raman

Foreword

By General Retired John Allen

I offer this foreword from the perspective of a general officer who served both as the Deputy Commander of the United States Central Command, and later as the Commander of the United States Armed Forces in Afghanistan. In those roles, I was either directly involved in the decision-making concerning or personally commanded the employment of drones in strikes against high-value targets in what is today being termed "drone warfare." Thus, I feel fully competent to write this foreword.

In this book, Paul Lushenko and Shyam Raman broaden our understanding of the perceived legitimacy of unmanned aerial vehicles, or so-called drones, on the modern battlefield. This evolving landscape is characterized by the emergence of novel technologies, some outfitted with artificial intelligence, that allow militaries to shorten the "sensor-to-shooter" timeline to maintain the lethal overmatch of adversaries while protecting their own forces. However appealing these advantages may be, they have raised important questions regarding the legality, morality, and ethics of employing drones during remote warfare.

One of the foundational questions relates to the perceived legitimacy, or appropriateness, of drone strikes. Over the course of my military career and public service, I have found that this question is both immensely important and poorly studied. Legitimacy is identified in U.S. military doctrine as a key joint warfighting principle, but it is unclear what makes drone strikes legitimate *per se*. While other experts have researched the relationship between public opinion and drones, nearly all of the studies I have read focus on attitudes of support and approval, mostly among Americans. Indeed, legitimacy is either mentioned in passing or used as a platitude, acknowledged as important for the sustainability of strikes abroad but rarely rigorously tested. To the extent scholars do attempt to understand what makes strikes legitimate or not in the eyes of the public, they mostly tap into the attitudes of U.S. citizens, though drones are increasingly used by other countries, whose citizens may—and often do, in my experience—perceive the appropriateness of strikes differently. As Lushenko and Raman state, failure to systematically study the public's— including those in other countries—mercurial understanding of legitimate drone strikes is puzzling, and it creates the potential for "blind spots" in the decision-making of U.S. political and military leaders.

In addressing this question, Lushenko and Raman advance the ongoing debate on public opinion and drones in important ways. First, Lushenko and Raman shed new light on an important social implication of drones by introducing and testing an original theory that allows scholars, policy-makers, practitioners, and industry leaders to understand how evolving patterns of drone warfare shape public perceptions of legitimacy. Rather than defining drone warfare in terms of the platform, targeted killing, or the much broader category of remote warfare, which is often the case, Lushenko and Raman conceptualize drone warfare as a function of observable and testable attributes, including why and how countries use strikes. Second, notwithstanding scholars' broad recognition that countries adopt different models of drone warfare, no one has systematically studied the implications of evolving patterns of drone strikes in the way Lushenko and Raman do. The authors draw on their original theory to conceptualize distinct models of strikes that allow them to explore the effects in terms of public perceptions of legitimacy and in a cross-national context. As far as I know, and based on my personal experiences as well as my own studies, they are the first to empirically validate unique American and French models of drone warfare, which are differentiated by the strategic and tactical purposes of strikes and degree of international oversight.

Finally, though scholars may recognize public perceptions of legitimacy as important to the sustainable use of strikes abroad, they neither adopt legitimacy as a dependent variable nor explore this social property using empirical data and statistical methods. Lushenko and Raman advance the research on public opinion and drones by explicitly using legitimacy as their dependent variable or main outcome of interest. They use original survey experiments to collect empirical data for public perceptions of legitimate drone warfare, given shifts in why and how strikes are used; use statistical methods to analyze the implications of evolving patterns of drone warfare for the public's perceptions of legitimacy in a cross-national setting; and identify instances when public support and perceptions of legitimacy coincide or deviate, thus exposing a "legitimacy paradox" that further justifies legitimacy as a meaningful consideration for scholars, policy-makers, practitioners, and industry leaders. This approach provides a new way of thinking about the implications of drones, given the broader strategic tapestry of conflict and war.

In my estimation, this book will appeal to a broad audience since it bridges the theory and practice of drone warfare in terms of the implications of strikes for public perceptions of legitimacy. Scholars will appreciate Lushenko and Raman's novel framework, which is informed by cross-disciplinary research and unique combat experiences, as a way to understand how the public adjudicates the legitimacy of drone warfare. Lushenko is perhaps the U.S. Army's foremost expert on drones and their implications for the shifting character of war. In a career spanning two decades, he has not only managed hundreds of

U.S. counterterrorism drone strikes across the world but written and spoken broadly on his experiences, renewing the literature on drone warfare and driving it forward in novel directions. Raman is an emerging star in public policy research, drawing on a wealth of experience supporting decision-makers in the U.S. government and military to help advance our understanding of the social impacts of emerging technologies in war.

Policy-makers, who often characterize strikes as legitimate to sustain their use abroad, will also appreciate the authors' theory as a means to assess the trade-offs for public opinion when using different models of drone warfare to balance vital national security interests against the enforcement of international norms. Practitioners, who routinely ponder the legitimacy of strikes they conduct on behalf of political officials, will also appreciate the analysis to forecast how variation in the use and constraint of drones can shape public attitudes that have important implications for the prospects of military success across the continuum of competition and in war. Those employed within the humanitarian and security sectors will similarly benefit from the book's dialogue with an emerging form of remote warfare that also shapes their professional obligations. Finally, drone manufacturers will find this book useful to baseline their understanding of the public's intuitions for legitimate strikes, which have implications for foreign military sales.

In sum, this book makes a critical contribution to our understanding of public attitudes toward drones, coinciding with an important inflection point in U.S. strategic culture for the place of emerging technologies on the battlefield, the legitimacy of which Lushenko and Raman help us better understand. Lushenko and Raman are on the leading edge of this research, which seeks to inform and shape policy at the intersection of emerging technologies and politics, and their analysis has implications in both domestic and global contexts. *The Legitimacy of Drone Warfare: Evaluating Public Perceptions* should be required reading in curricula, or in institutions, that seek to place the means of drones within the larger context of the ends, ways, and means of employing this technology in modern warfare.

1 Public Opinion and Drone Warfare

The U.S. counterterrorism strategy that emerged following the terrorist attacks on September 11, 2001, relied on armed and networked unmanned aerial vehicles (UAV), or drones, as they are now more commonly called, to kill suspected terrorists abroad (Kreps 2016; Kreuzer 2016). Unlike other capabilities that allow military forces operational stand off on the battlefield, such as artillery, bombers, and jets, drones have special qualities that are attractive to political officials. They are reusable, perform radical maneuvers, and provide persistent overwatch, which helps extend a military's operational reach (Boyle 2020). These characteristics also provide elected officials with a cost-effective way to achieve military objectives, such as removing high-value terrorists (Hardy & Lushenko 2012), while reducing risks to their own soldiers and safeguarding against unintended consequences, especially civilian casualties. While modern drones may have been initially exotic, their use outlasted then President George W. Bush, and their capability now constitutes a key feature of the United States' use of military force overseas. Indeed, despite varying degrees of transparency across successive presidential administrations—two Republican and two Democratic—since 2001, the U.S. drone program has enjoyed broad bipartisan support for over two decades (Kreps et al. 2022).

Bush's inaugural use of drones to kill terrorists also set a dangerous precedent for global security. Over 100 countries and many non-state actors have acquired drones, heightening the potential for conflict within and between countries (Rogers 2023b; Lushenko et al. 2022a). Together, these trends constitute what Agnes Callamard (2020), former United Nations (UN) Special Rapporteur on Extrajudicial, Summary, or Arbitrary Executions, refers to as a "second drone age" and have renewed the debate on drones. Among other topics, ranging from the proliferation to effectiveness to legality of drones, scholars continue to discuss the implications of public opinion for countries' use of strikes (e.g., Calcara et al. 2022a, 2022b; Schwartz et al. 2022; Chavez & Swed 2021; Rossiter 2018; Horowitz et al. 2016; Cortright et al. 2015). Though analysts have studied the relationship between public opinion and drone warfare at great length since 2001, the impact of public attitudes on

DOI: 10.4324/9781032614267-1

countries' drone policies lacks scholarly consensus, especially in terms of the public's perceptions of legitimate strikes.

This observation raises an important research question that constitutes the central focus of this book. How does the public perceive the legitimacy of countries' evolving use of drone warfare? Contrary to a legal-rational interpretation based on compliance with international law, namely International Humanitarian Law or the Laws of Armed Conflict (Blank 2023), we treat legitimacy as an empirical or pragmatic phenomenon. Legitimacy constitutes the subjective beliefs that the public has in the appropriateness of wartime conduct, including drone strikes. Among other scholars, Pan et al. (2022, 19) also contend that "legitimacy is a sociological phenomenon, and can only be meaningfully studied in the context of a society and the attitudes of individuals therein." What this assessment suggests, then, is that the concept of legitimacy is often used interchangeably with cognate terms, including attitudes of acceptance, appropriateness, and rightfulness. We also understand legitimacy in these semantic terms, especially because doing so allows us to better study the concept empirically, as we will discuss later.

These insights inform the main purpose of our book. We introduce and test an original middle-range theory that allows scholars, policy-makers, and practitioners to understand how evolving patterns of drone warfare shape the public's perceptions of legitimacy, which can moderate countries' drone policies and the global governance of drones. We argue that variation in countries' use of drones is best understood as a function of several observable—ergo, empirically testable—strike attributes. These include shifts in the use and constraints of drones to manage the potential for unintended consequences, namely civilian casualties. To the extent scholars interrogate evolving patterns of drone warfare globally, they mostly emphasize civilian casualties (Crawford 2003) or differentiate between the tactical and strategic uses of strikes (Chapa 2022; Brunstetter 2021; Boyle 2022). Few scholars recognize that countries can also constrain strikes differently through unilateral and multilateral measures, which we discuss in greater detail in the following chapters. This oversight is surprising given that security studies scholars increasingly assess the implications of restraint in civil wars and rebel governance, including the implications of landmine use on the legitimacy of rebel groups as perceived by both those afflicted by violence as well as member states of international society (Garbino 2023; Stanton 2016; Jo 2015).

In the context of drone warfare studies, the relative lack of attention to shifting constraints for strikes may belie an admonition by Carl von Clausewitz, the famous Prussian war theorist. He argued that such "restrictions" are "hardly worth mentioning" in the context of total war characterized by countries' large-scale mobilization of citizens and resources (Clausewitz 1984, 75). As we show throughout this book, drones have been used across the spectrum of war and during unique military operations, suggesting that they can be—and often are—constrained in different ways. Before introducing our research design and key findings for the public's perceptions of legitimate drone warfare, we need to know more about what the existing literature says about public attitudes toward strikes to help further justify our study.

Public Opinion and Drone Warfare

Writing in 1922, Lippmann defined public opinion as pictures people have in their heads regarding current events that they act upon when engaging political officials, which "governments find it prudent to heed" according to Key (1961, 14). Then, in 1992, Zaller showed that these pictures are a function of information that people act upon when engaging political officials, but only if they are disposed to do so. More recently, Kertzer and Zeitzoff (2017, 554) conducted empirical research, leading them to agree that "*individuals do carry substantively meaningful orientations toward foreign affairs around in their heads with them,*" which are the crux of public opinion. Given this understanding, the starting point for our empirical study is a bourgeoning body of literature that attempts to adjudicate public attitudes toward drones.

Kreps (2014) finds that while Americans generally support drone strikes abroad, the perceived compatibility of strikes with international law can moderate the magnitude of the effect. This is consistent with a study by Schneider and Macdonald (2016) that finds Americans support drone strikes when they are perceived to comply with domestic and international law. Kreps and Wallace (2016) further find that international and non-governmental organizations, such as the UN and Amnesty International, can shape public attitudes toward drone strikes, especially when criticisms relate to the legality versus effectiveness of operations. Horowitz (2016) also finds that public attitudes toward drone strikes are contextual, shifting based on the degree to which they protect soldiers. Similarly, Boddery and Klein (2021) find a successful drone strike can increase presidential approval despite a staggering economy, which evidently helps corroborate the diversionary use of force for domestic political gain that scholars have critically appraised in the past.[1] Finally, Lushenko and Kreps (2022) show that international approval is associated with greater perceived legitimacy for a drone strike, which mostly results from the assessed legality of operations.

While these and other studies benchmark how scholars think about public attitudes toward drone warfare, they also share several assumptions that suggest the need for further analysis. First, scholars generally draw on U.S. respondents to tap into public attitudes toward drone warfare. To be fair, this is an understandable practice. The United States is the most prolific user of drone strikes, meaning that Americans may serve as a useful barometer for global public opinion, which is also a key assumption in the literature for the international arms trade (Erickson 2018). The United States, for better or worse, also benchmarks democratic political regimes globally. The potential restraint afforded by public opinion on U.S. officials' use of drone strikes is thought to serve as an example for other presidential and parliamentary political systems (Reeves & Rogowski 2016; Watts 2008). Further, the data for U.S. drone strikes is better gathered and curated than it is for other countries, as well as drone strikes conducted by non-state actors, including al-Qaeda, the Islamic State, and the Houthis (Chavez & Swed 2023; Carter 2022; Schwartz et al. 2022). The Bureau of Investigative Journalism, New America, and other watchdog groups are designed to aggregate data for U.S. drone strikes. They attempt to account for America's "quasi-secretive" use of drone strikes abroad (Banka & Quinn 2018), which has resulted in more civilian casualties than U.S. officials publicly admit (Khan 2021).

Where existing databases rely on Americans' feedback, scholars face a crucial methodological trade-off in leveraging U.S.-centric information in their findings. There is increasing recognition that this approach does not account for selection bias, which undermines the generalizability of the results. Ceccoli and Bing (2018, 248) find "tepid (or at least mixed) support among European citizens presents a stark empirical contrast with such uniformly strong support in the United States" for drone strikes. In response, scholars increasingly administer cross-national survey experiments to gauge public opinion for drone warfare (e.g., Fisk et al. 2019). We follow suit by exploring how distinct models of drone warfare adopted by different countries can moderate public attitudes in unique ways, which we demonstrate in Chapter 3 through a novel study of French citizens' perceptions of legitimate drone strikes. Like the United States, France ranks among only a handful of great powers that use drones against terrorists abroad.

Second, while scholars assume that public opinion can shape officials' use of strikes, drones have unique qualities, making them a favored tool of elected officials, who often use strikes with impunity, apparently unencumbered by public preferences. In contrast to other indirect fire weapon systems such as artillery, bombers, and jets, drones erode reciprocal—physical—risk between combatants and remove targets' right to self-defense. Though Renic (2020) and other war ethicists (e.g., Enemark 2023) criticize these outcomes for imposing radically asymmetric violence that crosses over into morally problematic killing, it is precisely the anticipated savings for economic, military, and political costs that endear drones to officials, especially U.S. presidents (Blakeley 2021). Thus, it is possible to conceptualize drone warfare as a leader-driven practice that suggests public opinion is a feature of officials' decisions to use and constrain strikes in different ways (Lushenko 2022c; Payne 2020; Yarhi-Milo 2018). This helps inform our understanding of drone warfare as a function of shifting strike attributes that we discuss in the next chapter, wherein we introduce our conceptual framework. Though other scholars may agree, they are not so sanguine, casting especially U.S. presidents as "judge, jury, and executioner" when exercising drone strikes abroad (Sterio 2018).[2] While we do not stake out a position either way on presidents' use of drones, leaving such normative judgments to others, we do note that negative assessments are redolent of research that attempts to understand evolving presidential power for unilateral action. However, this research is based more on ambiguities embedded in the U.S. constitution than on the latitude seemingly provided by public opinion (Hathaway 2023; Moe & Howell 1999).

Western (2005, 225) finds that "Congress, like the public, has tended to defer to the judgment of the executive branch, and popular presidents are given more latitude" to use force abroad, which Canes-Wrone et al. (2008) empirically show to be the case. Horowitz et al. (2015, 55) contend that this suggests a broader trend in the "foreign powers of the U.S. president" that "have been increasing for several generations." Representative Steve Israel (D-NY), who served on both the Armed Services Committee and Appropriations Subcommittee for Defense, agrees. He recalls that Congress "did not effectuate sufficient oversight related to the drone program" during his time in office, which spanned two decades following the terrorist attacks of 9/11 (Israel 2022). This, despite protests from other congressional officials like Senator Ron Wyden (D-OR) that presidents "should not be allowed to conduct such a serious and far-reaching program by themselves without any scrutiny because that's not how American democracy works" (Mapes 2013), which is a sentiment echoed by other congressional representatives as well (Müller & Böller 2022).

Third, scholars have not studied, at length, "microfoundations" of citizens' preferences for drone strikes. According to Kertzer (2017, 83), microfoundations help explain "outcomes at the aggregate level via dynamics at a lower level." Researchers' inattention to microfoundations may reflect a shared belief among scholars that public attitudes are uniquely structured in terms of drone strikes (Hurwitz & Peffley 1987). Ceccoli and Bing (2018, 248) argue that "ideology and core policy beliefs shape respondent sentiment toward drone strikes in clear and convincing ways." Yet these and other scholars do not theorize about the mechanisms that may underlie public attitudes toward drone strikes, whether they relate to "cold" (cognitive) or "hot" (affective) processes (Kertzer & Tingley 2018). Using a survey experiment of American respondents, Kreps and Wallace (2016) conclude that public opinion for strikes seems to relate more to normative (e.g., moral and legal) concerns than simply instrumental (e.g., national security) interests. While this echoes recent scholarship for the United States' use of force more generally (Dill & Schubiger 2021), Kreps and Wallace (2016) caution that their findings deserve more study to elucidate the exact values or beliefs that explain this observation, which other scholars similarly encourage (e.g., Fang & Oestman 2022; Sagan & Valentino 2018; Hazelton 2017).

Where scholars have explored the microfoundations of the public's attitudes toward drones, they often use causal mediation analysis to show the complete causal chain for the effect of an independent variable (e.g., international approval) on a mediator (e.g., political ideology) and the effect of a mediator on a dependent variable (e.g., approval and support) (e.g., Horowitz & Lin-Greenberg 2022; Lin-Greenberg 2022; Fang & Oestman 2022; Imai et al.

2011). Fisk et al. (2019) attempt to isolate the microfoundations that may shape public opinion for drones, choosing to study the implications of anger and fear for the public's support for strikes in a cross-national—France, Turkey, and the United States—context. The results are mixed. The authors report statistically significant evidence for the mediating effects of anger on public support for drone strikes across all three countries, but null effects for fear.

While the authors' use of causal mediation analysis is also novel, though not without criticism, as we explain in Chapter 3, the theoretical intuition for emotions as potential mediators of public attitudes is difficult to justify. Substantial research shows that at least Americans are ambivalent about strikes abroad (Kreps 2016), and U.S. soldiers are even more clinical in their assessments, reflecting little to no emotions when adjudicating their support for strikes (Lushenko et al. 2023). Even if we assume that Americans are abreast of officials' use of drone strikes abroad, research shows that they prefer to outsource their accountability for strikes to others, including Congress and watchdog organizations like the Bureau of Investigative Journalism and New America, unless or until strikes affect them, kill U.S. citizens, or result in civilian casualties. According to Dudziak (2012, 8), for instance, "we see occasional headlines about drone attacks . . . but war has drifted to the margins of American politics."[3]

Finally, public opinion researchers invariably adopt approval or support as the outcome variables of interest in their studies. They share the assumption, mostly implicit, that these attitudes primarily define citizens' engagements with political officials' decisions to use drones. In doing so, scholars discount other attitudes that could also influence public preferences for drones, such as reputation for resolve[4] (e.g., Lin-Greenberg 2022; Kertzer 2016; Mercer 1996) and perceptions of legitimacy (Lushenko 2022a). Legitimacy is not only important in its own right, as we discuss in the next chapter. But it also appears to have importance as the *sine qua non* of success in "new wars" against non-state actors, including the use of drone strikes against insurgents and terrorists (Kaldor 2018; Dill 2015). Yet Lewis and Vavrichek (2016, 172) caution that there has been an "inadequate consideration of *legitimacy*" in countries' drone policy and scholarship. This echoes Regan (personal communication, June 5, 2022), who avers that "there has been little effort to systematically study legitimacy" in terms of drone warfare.

The failure of scholars to empirically study legitimacy in the context of drone warfare is puzzling. Research suggests that the public's attitudes of approval or support are not synonymous with its perceptions of legitimacy. In tracing shifts in coalition warfare since the Cold War, for example, Kreps (2011) finds that the public's perception of legitimacy for multinational operations "becomes a worthy end in itself, even if it introduces inconveniences along the way." In the context of autonomous weapons, or those who can identify, track, and prosecute targets on their own, Ferl (2023) finds that competing visions over human oversight contribute to the perceived legitimacy of these emerging capabilities. While political officials and scholars also ac-knowledge that the public's perceptions of legitimacy are a core feature of drone strikes, few experts investigate legitimacy empirically, creating the po-tential for biased judgments (e.g., Regan 2022; McDonald 2021; Barela 2015; Dill 2015). On the one hand, scholars often speak in terms of the legitimacy of drone warfare. McDonald (2021, 539), for instance, claims that legitimacy is "central" to countries' adoption of drones, which echoes a recent finding that perceptions of legitimacy are decisive in shaping public attitudes toward military interventions (Fang & Oestman 2022). On the other hand, U.S. of-ficials, who are the most prolific users of drones globally, characterize their strikes as "righteous" (Aikins et al. 2021), even when they inadvertently kill civilians. In doing so, "U.S. officials routinely invoke polling data to enhance the legitimacy of their policy actions" (Rowling & Blauwkamp 2021, 16), including drone strikes.

This process of legitimation (Goddard & Krebs 2015) is designed to preserve the longevity of the U.S. drone program, which often lacks transparency (Banka & Quinn 2018), can contravene international law (Kaag & Kreps 2014), and re-sults in more collateral damage, especially civilian casualties, than is publicly acknowledged (Lushenko et al. 2022). On May 3, 2023, for instance, the U.S. Central Command killed Lotfi Hassan Misto in northwest Syria with a drone strike, though no evidence exists that this elderly sheep herder was connected to terrorists, as U.S. military officials claimed (Nezhat et al. 2023). Former Direc-tor of the U.S. Central Intelligence Agency, General (Retired) Michael Hayden, notes that "no president can do something repeatedly over a long term without the broad public support" (Hopkins 2013). Indeed, research shows that the pub-lic does not typically challenge actions it perceives as legitimate (Voeten 2005). It is therefore unsurprising that Pong (2022, 382) finds that public opinion, in-cluding perceptions of legitimacy, "is key in normalizing, as well as garnering funding for, military autonomous systems" such as drones. At the same time, Hodges (2018, 17) shows that "militaries in a democracy are concerned that their actions must be perceived as legitimate, since to lose legitimacy may well undermine the realization of strategic benefits."

In sum, legitimacy is a consequential public attitude that can shape policy, especially in terms of the use of force abroad. It is not analogous to other dispo-sitions, though scholars often, perhaps unintentionally, conflate legitimacy with other attitudes, particularly approval or support. Rather, as Blank (2023, 266) argues, legitimacy is a separate and "essential component of any military opera-tion," particularly through drones given their heightened proliferation globally.

Studying Public Perceptions of Legitimate Drone Warfare

Some may contend that a focus on the public's perceptions of legitimacy is too stylized, meaning that the concept is insufficiently different from other attitudes and possesses little explanatory and predictive value for countries' evolving use of drone strikes. Rather, legitimacy is in the eye of the beholder and difficult to generalize, especially across time and space. Yet we lack systematically derived evidence to substantiate these unqualified claims. At most, scholars note that legitimacy claims for the use of force abroad, including drone strikes, are a matter of judgment (Schmidt & Williams 2023). This suggests that there is a gap in the quantitative research agenda for public attitudes toward drone warfare in terms of understanding perceptions of legitimacy. One leading advocate for the study of how the public understands legitimate drone warfare argues that "[t]he fact that human judgment is unavoidable does not mean that all analysis is simply the reflection of subjective preferences that cannot be subject to rigorous assessment" (Regan 2022, 12).

How, then, should we study the public's perceptions of legitimate drone warfare? Students of International Relations (IR) have long adopted a "classical" approach to adjudicate the public's perceptions of legitimacy, which is to say, through reasoned judgment (Dunne 1998; George 1976). One cardinal advocate of this methodological approach, Hedley Bull, argued that its value "consists essentially of judgements that are not established by mathematical or scientific methods" and "may be arrived at quite independently of them" (Bull 1966, 368). "Such exercises of judgement are not arbitrary or mere appeals to intuition or assumed authority," Bull contended. He offered that "they are verified or falsified by examination of the world" (Bull 1975, 279). War ethicists such as Christian Nikolaus Braun agree. Braun, along with other neoclassical just war thinkers, adopts a method of casuistry to "derive judgments about the rightfulness or wrongfulness of action by investigating particular cases" (Braun 2023, 56).

On the other hand, positivist political scientists, or those who attempt to demonstrate a causal association—if not relationship—between different social phenomena, recognize that it is also possible to study legitimacy with empirically derived data and while using statistical or econometric techniques (Dellmuth & Tallberg 2023; Fang & Oestman 2022; Jongen & Scholte 2022; Pan et al. 2022; Binder & Heupel 2021; Tallberg & Zurn 2019; Mikhail 2009). This quantitative approach, ironically, is informed by the observation of another classical theorist, David Beetham. He (1991, 13) contented "evidence [of legitimacy] is available in the public sphere, not in the private recesses of people's minds."

In this book, we capitalize on this observation to study the public's perception of legitimate drone warfare with multiple methods and in different strategic contexts (Pearl & Mackenzie 2018). Specifically, we adopt a quantitative research design consisting of survey experiments and descriptive and inferential statistics, including multivariate regression analysis, simulations, and causal mediation analysis.[5] In the next chapter, we explicitly define drone warfare, legitimacy, and the relationship between these two concepts. While scholars often reference these contestable concepts, they rarely define them as well as their intersection. We also present our original conceptual framework, which we operationalize in subsequent empirical chapters to adjudicate the implications of evolving patterns of drone warfare globally on the public's perceptions of legitimacy.

In Chapter 3, we draw on this original middle-range theory to derive testable hypotheses for the implications of evolving patterns of drone warfare on the public's perceptions of legitimacy in a comparative context. We posit that citizens prefer distinct models of drone warfare that differ according to how countries' use and constrain strikes to help mitigate unintended consequences. To test this intuition, we conducted original survey experiments among respondents in two great powers that frequently use drones for counterterrorism beyond their borders and regions: France and the United States. Our samples are nationally representative and total over 1,800 respondents. The results offer strong statistical support for our theoretical expectations. French respondents prefer what we refer to as "juridical" strikes, which consist of the tactical use of drones with multilateral constraints. Moreover, they endorse this pattern unconditionally or regardless of the potential for civilian casualties. American respondents, on the other hand, prefer what we call "over-the-horizon" strikes that relate to the strategic use of drones with unilateral constraints. We also find that the type of constraint Americans endorse is conditional on the potential for civilian casualties. We also show that Americans and French respondents can empirically identify their preferred pattern of strikes when presented with randomized and sanitized scenarios that do not identify the targeting country, which provides the first direct evidence of cross-nationally distinct models of drone warfare.

Importantly, we also find that while public attitudes of support and legitimacy can coincide in terms of different patterns of strikes, they can also deviate, suggesting a "legitimacy paradox" in which the public may perceive strikes as more or less legitimate compared to their level of support. This is similar to what other scholars call a "trust paradox" in terms of artificial intelligence (AI), whereby people may support AI-enabled technologies but not trust them as much, such as driverless vehicles (Horowitz et al. 2023; Kreps et al. 2023). Our finding of a "legitimacy paradox" in terms of drone warfare justifies our use of causal mediation analysis to help explain why this is the case. We show that French citizens' understanding of legitimacy is shaped by the perceived morality of strikes, a belief in the role of great powers, and a preference for the use of force abroad. Americans' understanding of legitimacy, meanwhile, is largely shaped by the perceived legality of strikes, which is a puzzling outcome considering U.S. drone operations by definition often breach other countries' sovereignty or territorial integrity in the pursuit of terrorists abroad. We explain what this might suggest about U.S. preferences for drone warfare.

In Chapter 4, we explore the implications of the "certainty" standard for the civilian casualty outcomes following U.S. drone strikes in Pakistan, which is one key type of unilateral constraint that scholars have largely eschewed.

In addition to reconciling this oversight, our focus is based on findings from Chapter 3 that suggest that U.S. citizens' perceptions of legitimate drone warfare can be moderated by the unintended consequences, namely civilian casualties. This outcome is shaped, in large part, by the type of constraint.

We exploit a dramatic shift in the U.S. drone policy from 2011 to 2013 to adjudicate the merits of the near certainty standard. Contrary to the more permissive reasonable certainty standard adopted by Bush and then Trump, Obama conditioned strike approval on the near certainty of no civilian casualties. For the first time in the literature, our statistical analysis, coupled with rare interviews with senior Obama-era officials, shows that the policy was actually adopted in 18 to 30 months prior to Obama's official announcement at the National Defense University in Washington, DC, on May 23, 2013, which scholars usually define as the official implementation date. This new finding is integral to better understanding the policy's effectiveness over time and has important implications for the public's perceptions of legitimate strikes. Indeed, we find that the more stringent targeting standard prevented the deaths of hundreds of Pakistani civilians, averted a significant social and economic toll across Pakistan, and came at no appreciable cost to U.S. national security. Thus, we empirically validate Obama's earlier intuition that he instituted a policy "least likely to result in the loss of innocent life" (Obama 2013). Importantly, the existing field research suggests that the near certainty standard succeeded at enhancing Pakistani citizens' perceptions of the legitimacy of U.S. drone strikes (e.g., Mahmood & Jetter 2023; Ansari 2022a; Shah 2018; Williams 2011).

In Chapter 5, we explore the implications of one key type of multilateral constraint, international approval through the UN, which is also based on findings from Chapter 3 that suggest that French citizens' perceptions of legitimate drone warfare can be significantly moderated by how strikes are managed. To better understand this dynamic, especially because it provides further leverage over the relationship between public opinion and emerging models of drone warfare globally, we conducted survey experiments across nationally representative samples in France and the United States, again totaling over 1,800 respondents. Our results reflect that international approval heightens the perceived legitimacy of a strike, especially among French citizens, which further corroborates their preferred model of juridical strikes. Also echoing findings in Chapter 3, we show that Americans emphasize international law the most as the basis of their perceptions of legitimacy, even though their preferred model of over-the-horizon strikes often transgresses other countries' territorial integrity. In contrast to French citizens, we also find that Americans emphasize burden-sharing among allies and partners in terms of strikes. Together, these findings show that while Americans may want their leaders to conduct strikes in partnership with other countries, they generally perceive the legitimacy of over-the-horizon strikes as greater, seemingly because it preserves the United States' freedom of action globally. This model of drone warfare, in other words, does not tie the hands of officials in using force abroad, including when, where, and how they like. We recapitulate our findings in the concluding chapter while discussing the possible limitations of our analysis and suggesting a future research agenda.

Contributions and Limitations

Overall, these findings make important theoretical and empirical contributions to the research agenda for the relationship between public opinion and drone warfare. First, understanding drone warfare in terms of observable strike attributes, including the use and constraint of strikes designed to help minimize unintended consequences, especially civilian casualties, provides a generative framework to interpret evolving patterns of drone warfare globally. As we discuss in the next chapter, scholars often emphasize the "agentic capacity of drones" (Demmers & Gould 2020), conceptualize drone warfare in terms of targeted killing only, and often fail to appreciate that countries can vary both why and how they use strikes within and across their borders. Our novel middle-range theory provides further leverage to explore public opinion for drone warfare in emerging strategic contexts, particularly the public's perceptions of legitimacy, which can have implications for countries' drone policies. This is especially acute in democratic political regimes, which are prefigured on elected officials' accountability for the use of force abroad. Thus, we also respond to previous research that commends scholars to "unpack" the mechanisms of perceived legitimacy in the context of military interventions abroad, of which drones are now a favored tool (Fang & Oestman 2022).

Second, we push the methodological boundary by showing that legitimacy can be a useful dependent variable for empirical researchers, like it is for political theorists and "classical" IR scholars, in gauging public attitudes toward the use of force abroad. Though scholars, policy-makers, and practitioners understand legitimacy as a critical condition for the sustainability of drone strikes, this outcome is generally ignored by mainstream public opinion researchers or those who draw on empirical data to adjudicate citizens' preferences for military operations. Similar to public attitudes in terms of approval or support, we then require a better understanding of the public's subjective beliefs about the appropriateness of countries' wartime conduct.

We confront this task in the context of drones by building on existing studies on the legitimacy of strikes, which almost always use qualitative evidence and methods, meaning their findings can be non-falsifiable, difficult to replicate, and hard to generalize (e.g., Blank 2023; Dorsey & Amaral 2021; Barela 2015; Dill 2015). We collect data for the public's perceptions of legitimate drone warfare given shifts in the use and constraint of strikes to help mitigate civilian casualties. We then use statistical techniques to analyze the implications of these evolving patterns of strikes for the public's perceptions of legitimacy in a cross-national setting while also providing an initial probe of the implications of shifting—unilateral and multilateral—constraints on legitimacy outcomes for strikes among two countries that frequently use drones for counterterrorism beyond their borders and regions: France and the United States. We find that legitimacy matters for observers of drone strikes; evolving patterns of strikes can shape legitimacy outcomes in different and predictable ways cross-nationally; and that while perceptions of legitimacy often align with attitudes of support, they can also diverge, suggesting a "legitimacy paradox" that may have important implications for policy.

Notwithstanding these core contributions, we recognize that our study is also limited in multiple ways. First, as with all survey research that does not use longitudinal data to study shifts in public attitudes over time, we do not examine historical—social, economic, and political—factors that may also shape perceptions of legitimate drone warfare among Americans and French citizens. In other words, our study, while novel, only provides a cross-sectional understanding of the public's perceptions of legitimate drone warfare that scholars can build on. We provide broad attitudinal trends given evolving strike profiles, but at the expense of a greater understanding of public attitudes in unique cases, such as the use of drones for political assassination. Second, the extent to which our findings relate to different political regimes involved in different types of conflict beyond counterterrorism and while using smaller, cheaper, commercially available, and easily weaponized drones is also unclear. Kunertova (2023, 1), for instance, observes that the "game-changing effect of drones depends on the game." As multiple analysts further note (Chavez & Swed 2023; Chavez 2023; Cronin 2020), in the game of dual-use capabilities, the security challenges are not understood as well by scholars. Indeed, the sales of hobbyist drones globally are projected to grow to $43 million by 2024, which constitutes over a 300% increase since 2018 (Drone Industry Insights 2021). The proliferation of these capabilities is likely to have important implications for the public's perceptions of legitimate strikes that our middle-range theory can neither predict nor explain.

Third, our study is also limited by a focus on attitudes drawn from the publics of targeting versus targeted countries and for a discrete set of circumstances. Notwithstanding concerns for selection bias and reverse causation, there is increasing recognition that attitudes among those targeted by drones may be more important for the sustainability of political officials' use of drone strikes abroad and that they can vary due to different conditioning variables (Page & Williams 2022). Finally, critics may question assumptions underlying our statistical methods that have important implications for the durability of our findings in different countries, among different audiences, and at different periods of time. Fortunately, our research provides an important baseline to address these and other concerns that seem increasingly pressing given a renewal of great power competition that is exacerbated by the proliferation of drones globally, which has heightened violence within and between countries and possibly to the point of shifting the offense–defense balance between countries during war (Calcara et al. 2022b; Lushenko & Kreps 2023b). In the next chapter, therefore, we clarify key concepts in the interest of introducing a middle-range theory that provides a parsimonious yet generalizable way to understand how evolving patterns of drone warfare shape the public's perceptions of legitimate strikes.

Notes

1 On diversionary use of force, see Meernik and Waterman (1996).
2 Stein also argues that countries "cannot think, process information, estimate probabilities, or calculate, only their leaders can" (Stein 2017, S256).
3 See also Kaag and Kreps (2014).
4 According to Kertzer (2016), resolve is a function of dispositional and situational considerations that reflect the extent to which an actor maintains an intention despite the temptation to back down during conflict.
5 The data that supports the findings of this book are available from the corresponding author, Paul Lushenko, upon reasonable request.

2 Drone Warfare, Legitimacy, and Global Patterns of Strikes

Drones have occupied the battlefield for centuries and have their origins in the ancient world when humans learned to "throw fire" (Crosby 2002). Observation balloons, such as those used during the American Civil War (1861–5), represented a primitive version of a drone. The merger of projectile technology with the airplane in the early 20th century ultimately led to the emergence of modern armed and networked drones, as multiple scholars show (e.g., Rogers 2023a). The development of weaponized drones over time was never so linear, however. Military innovation, especially in the United States, often progressed in fits and starts, subject to competing missions and cultures across the armed forces that shaped uneven research, development, procurement, and sustainment of drones from World War II until 9/11 (Lee 2023; Jensen et al. 2022; DeVore 2020). Set against this context, some airpower theorists argue that modern drones constitute a "Revolution in Military Affairs" that alters the character, if not nature, of war. What this implies is that drones not only affect how countries fight but also why they fight. Among drone revolution proponents (Calcara et al. 2023), James Rogers is most outspoken, arguing that drones have been the most significant development ever in weaponry, poised to "help decide the fate of nations" (Rogers 2020, 41; Callamard & Rogers 2020).

Yet the revolutionary status of drones is a highly contestable topic, exacerbated by the integration of these capabilities into interstate wars, such as the conflicts between Armenia and Azerbaijan, as well as in Libya and Ukraine. Some experts suggest that drones provide a decisive advantage to countries on the battlefield and may even threaten to undermine the liberal global order (Lushenko et al. 2022). Other experts caution that drones are "highly vulnerable to air defenses," implying they do little to shift the offense–defense balance between countries during war or shape overall war outcomes (Calcara et al. 2022b, 2022c). This debate, while useful, often puts the cart before the horse. Scholars do not clearly define what drone warfare is and is not, which is integral to our study of the public's perceptions of legitimate strikes.

The purpose of this chapter is to define key concepts that frame our original middle-range theory. We achieve this aim in three parts. First, we offer an

DOI: 10.4324/9781032614267-2

original definition of drone warfare, which we argue is best understood as a function of observable and empirically testable strike attributes. Second, we relate this definition to legitimacy. Finally, we draw on these advancements to introduce a novel middle-range theory that helps us predict and explain variations in the public's perceptions of legitimate drone warfare. In doing so, we provide a parsimonious yet generalizable way to understand public perceptions of legitimate strikes in cross-national contexts, especially among democracies that are prefigured on elected officials' accountability for their use of force.

Drone Warfare

While scholars often refer to so-called drone warfare, they do not explicitly define this term. Drone warfare may be an intuitive concept. But it is also a loaded concept. Critics often use the term as a euphemism to expose the arguably unethical, immoral, and illegal dimensions of strikes based on the respatialization of war, namely pilots' disconnection from the battlefield (Chamayou 2013). A review of the literature suggests that scholars define drone warfare in one of four main ways, imposing key trade-offs for understanding the intended purpose of strikes and the implications for public opinion.

First, researchers often conflate drone warfare with the unmanned aerial vehicle (UAV) platform itself, resulting in what Gusterson (2019) calls "drone essentialism." Drones, in other words, represent what Biswas (2014, 110) refers to as "a fetishized object of state desire." This perspective fails to appreciate that drones are merely one component of a globe-spanning intelligence collection and analysis architecture that U.S. Air Force Colonel Michael Kreuzer (2016) refers to as a "system of systems."[1] It can take more than 200 personnel to operate a drone for 24 hours, and over a quarter of these enablers are deployed within a theater of operations or to expeditionary bases close to a conflict (Benjamin 2012). Second, scholars often equate drone warfare to remote warfare, though the latter concept is broader and incorporates a range of stand off capabilities, including cyber, drones, and proxy forces (e.g., Renic 2023a; Biegon et al. 2021; DeShaw Rae 2014). Third, experts often conflate drone warfare with the targeted killing of terrorists, thereby discounting the use of drones for other purposes. The most insightful analysis recognizes that armed and networked drones can "kill," "watch," or "aid" (Welsh 2015).[2] That is, drones can enable humanitarian assistance and disaster relief, perform surveillance and reconnaissance, facilitate preparatory fires prior to ground missions, and provide close air support to ground forces, among other missions.

Finally, scholars do not define "warfare" in drone warfare, meaning the term is often conceptually stretched. Though Fearon (1995) cautions that the scholarship on war is "massive," Lyall and Wilson III (2009, 71) caution that a failure to define war "is problematic if the determinants of outcomes vary systematically by conflict type and time period." Indeed, scholars researching drones have privileged several definitions of war that reflect the broader literature on the causes and consequences of conflict. Some analysts interpret drones themselves as structuring war. Renic (2020, 159), for instance, understands "UAVs *as* war, rather than *in* war." Other scholars adopt a social-psychological perspective that relates war to leaders' predispositions. In this case, leaders are thought to adopt unique cognitive frames or worldviews that shape their understanding and use of drone strikes (Lushenko 2022c). Still, other experts interpret war as an ineradicable human condition, akin to *animus dominandi* (the desire for power), that is thought to drive countries' international relations. In the context of drones, Kaag and Kreps (2014, 17) argue their banality has encouraged a "moral hazard" wherein officials "have an incentive to undertake the risky behavior of targeted killing because technology shields them from most of the adverse consequences." Another cohort of researchers also assumes that countries only use drones during irregular warfare against terrorists. The opposite is also true, as the war in Ukraine reflects, meaning drones are increasingly used during interstate conflict (Rossiter 2023; Hamming 2023). Indeed, a final stable of experts understands war in a "strict" sense, which results from the large-scale mobilization of countries' armed forces, meaning that while drones may be necessary to help achieve political and military objectives, they are insufficient on their own to do so during high-intensity combat (Calcara et al. 2022b, 2022c).

Some researchers reconcile this definitional pitfall by differentiating between varying strike profiles or how countries use strikes. Gusterson (2015) conceives of "pure" and "mixed" strikes, noting drones can be used separate from or in support of deployed forces. Chapa distinguishes between "tactical" and "strategic" strikes; this is a distinction that Boyle (2020) draws as well.[3] Commanders can use drones during an engagement with combatants, while political officials can use drones as a "foreign policy tool" (Chapa 2022).[4] Brunstetter (2021) bands strikes against "self-defense" and "cooperation" types. The former consists of drones used to kill terrorists. The latter relates to drones used during internationally approved interventions. Braun (2023) further conceptualizes drone strikes as either retributive or anticipatory, which are designed to impose punishment for some grievance or prevent some forecasted threat from transpiring.

Less attention is paid to how countries can also constrain strikes (Haun 2022), despite Galliott's (2015, 4) contention that while political officials may have "an obligation to utilize unmanned systems, their deployment should be subject to strict oversight." This is because, as Challans (2007, 154) contends and in contrast to Clausewitz, "[w]ith restraint comes legitimacy." Bain (2023) and Enemark (2023) go further and characterize constraint as a "virtue," meaning that the practice of restraining the use of force abroad has moral purchase in global politics. Thus, we argue that drone warfare is best understood in terms of how countries combine different—tactical or strategic—uses of strikes with varying—unilateral or multilateral—constraints to help prevent unintended consequences, especially civilian casualties.

Legitimacy

At the same time, scholars, policy-makers, and practitioners reference the legitimacy of drone strikes, though this term is poorly defined and rarely, if ever, empirically tested. To be clear, legitimacy is difficult to measure and is often excoriated as non-falsifiable. Lake counsels that any conception of legitimacy is dubious because scholars cannot hold social conditions constant (Lake 2011). Verba (1971), Hodges (2018), and Luft (2020) also note that this is a real concern when attempting to generalize empirical findings in a cross-national or comparative setting, especially considering cultural differences across societies can also shape public perceptions of appropriate behavior. As we addressed in the introductory chapter, however, scholars' decision to eschew the public's perceptions of legitimate drone warfare is puzzling, and more so because classical IR theorists recognized the importance of legitimacy for the durability of global politics.

Writing in the 17th century, for example, Hobbes (1994) contended that legitimacy is foundational to prosocial behavior between political communities as well as enduring peace. That legitimacy does not hold the same cache for scholars nowadays, at least as a dependent variable in drone warfare studies, may be a function of deeper theoretical assumptions about countries' foreign relations. Whereas classical realists took legitimacy seriously (Morgenthau 1948), structural realists believe legitimacy is a vacuous concept in global politics (Gelpi 2003).[5] The reality is, of course, more complicated. At stake is the relationship between legitimacy and power, defined by the rationalist literature as material capabilities that afford strong countries leverage over weaker ones (Drezner 2021). Neo-realists conceive of these two concepts as contradictory. At most, they attribute legitimacy to the preferences of great powers that attempt to capitalize on the promised dividends of perceptions of legitimacy, namely acceptance of paternalistic and predatory behavior (Hurrell 2004). Legitimacy for neo-liberal institutionalists is also problematic. They claim ideational factors can "mediate the relationship between material and other interests and political outcomes" (Reus-Smit & Snidal 2010, 21). In the same breath, they also maintain that "norm-following behavior is challenged by the logics of strategic interaction and strategic rationality" (Keohane 2010, 3). Power and interests come first for realists and neo-liberal institutionalists, implying that the circle is difficult to square for the place of legitimacy in global politics. Or, as Goddard and Krebs (2015, 6) put it, legitimacy is little more than "window-dressing for interests and power."

Constructivists conceive of anarchy, defined as the lack of a supranational authority to order countries' behaviors, in terms of culture, suggesting that social institutions such as legitimacy are enmeshed with actors' international relations (Wendt 1992). Wendt (1999) argues that legitimacy follows when "an actor fully accepts its claim on himself." This coheres with the so-called English School, which is predicated on the possibility of an international society of states bounded by common norms, rules, and institutions that can overcome the travails of anarchy, namely self-help behavior that threatens a security dilemma, heightened enmity, and a spiral toward war (e.g., Jervis 2017; Dunne 1998; Bull 1977). Clark, for example, interprets legitimacy as a function of norms—defined as standards of behavior for actors of a given identity—that inform the public's expectations of rightful conduct and membership in international society (e.g., Clark 2005; Katzenstein 1996).

These inconsistencies mean that defining legitimacy is vexing. Some scholars understand legitimacy sociologically or on the basis of "rules of the game" that prescribe or proscribe behavior (Bukovansky 2002). Others define legitimacy normatively, which is to say, based on certain expectations of behavior. The public may consider actions "desirable, proper, or appropriate within some socially constituted system of norms, values, beliefs, and definitions" (Suchman 1995, 574). Still others define legitimacy empirically. Legitimacy is what the public believes is acceptable (Weber 1968). These divisions have caused theorists to further bin legitimacy into "subjective" and "objective" buckets (e.g., Mittiga 2022; Buchanan & Keohane 2006). The latter is a matter of compliance with international law, namely International Humanitarian Law. The former is shaped by personal preferences.

Such typologizing is disorienting. More importantly, it may draw a distinction without a real difference. Rules of the game, regardless of their informality, reflect certain norms that shape the public's expectations of rightful wartime conduct, which now include countries' use of drone strikes abroad (e.g., Dill et al. 2022; Lushenko 2022a; Nichols 2018; Krasner 1983). These expectations are empirical in the sense that they are "what people accept because of some normative understanding or process of persuasion" (Hurrell 2004, 16). In the context of drones, we contend that legitimacy constitutes the subjective beliefs that the public has in the appropriateness of strikes based on how they are used and constrained to help minimize civilian casualties, which emerged as a legal, moral, and strategic imperative following World War II (Fazal 2018).

Some may criticize this definition as bordering on tautology (Grafstein 1981). Legitimacy, in other words, is in the eye of the beholder and, therefore, dependent on the context of drone use. Yet our descriptive definition offers at least two key advantages for the purpose of causal inference, or determining precisely how shifting patterns of drone warfare shape the public's perceptions of legitimate strikes. First, it provides analytical precision. That is, it has the benefit of transcending an important but altogether separate consideration surrounding the normative merits of countries' use of strikes, or if countries ought to use drones abroad assuming they can in the first place. Stated differently, we leave the prudence of countries' use of drones, which Morgenthau (1948, 12) called "the supreme virtue of global politics," to other scholars privileging a classical approach to IR. Aslam, for instance, incorporates prudence into a normative framework to help intuit the implications of drones for the future of U.S. foreign policy, affording us the ability to explore how people may perceive the legitimacy of these operations in more practical-theoretical or descriptive terms (Aslam 2013). Second, contrary to researchers who protest legitimacy is too "slippery" to study empirically, our descriptive understanding of legitimacy makes the concept measurable, testable, and falsifiable, especially with the experimental approach and quantitative methods we adopt (e.g., Pan et al. 2022; Hurrell 2004). Integral to our empirical analysis is specifying the two mechanisms or attributes that we hypothesize can shape the public's perceptions of legitimate strikes in cross-national contexts.

Global Patterns of Drone Warfare

Countries' adoption of drones varies in terms of why and how they are used. Countries can use strikes tactically or strategically, which differ by their intentionality and location (Table 2.1). Countries often conduct strikes in declared and undeclared theaters of operations. Contrary to active conflict zones, such as Afghanistan and Iraq, non-active conflict zones, including Pakistan, Somalia, and Yemen, are characterized by operations that are neither sanctioned by the UN nor have intervening countries' forces deployed on the ground. Countries also govern strikes through unilateral or multilateral constraints (Table 2.2). These differ according to who or what holds countries' use of strikes accountable to key standards of wartime conduct, namely the *jus in bello* (justice in war) principle of distinction or non-combatant immunity. Indeed, countries often flout *jus ad bellum* (justice of war) norms, such as just cause and proper authority, while also attempting to fulfill *jus in bello* norms.

Use—Tactical or Strategic

Countries use drones during discrete engagements with combatants to achieve near-term and limited military objectives in declared or "hot" theaters of operations (Boyle 2020; Cook 2015). The tactical use of drones suggests several observable indicators. Commanders use strikes during engagements in support of ground forces. According to Phelps, this characterizes a majority of strikes. Drones loiter above conflict zones, waiting to identify "someone to be killed or something to be destroyed" (Phelps 2021, 72).[6] Boyle adds that the "role that reconnaissance Predators and Reapers played in the hunt for IEDs [Improvised Explosive Devices] that were killing and wounding US troops amid the insurgency in Iraq shows how they could have direct tactical value" (Boyle 2020, 108). Drones are also deployed on an expeditionary basis, meaning strikes are mission-oriented. As such, strikes do not typically contravene the sovereignty or territorial integrity of countries, especially because they are used in the context of internationally recognized conflicts.

Countries also use drones strategically (Biegon & Watts 2022; Cook 2015). The strategic use of drones is more comprehensive, deliberately planned, and

Table 2.1 Varying Use of Drone Strikes

Use	Location	Execution	Purpose	Observable Implications
Tactical	Declared theaters of operations	Discrete and hasty	Near-term and limited military objective	• Commanders use • Supports ground forces (reciprocal risk) • Expeditionary deployment of drones • Respect for sovereignty
Strategic	Undeclared theaters of operations	Comprehensive and deliberate	Long-term and broader military and/or political objective	• Executive officials use • Not in support of ground forces (no reciprocal risk) • Global network of bases • Erosion of sovereignty

Table 2.2 Varying Constraints of Drone Strikes

Constraint	Scope	Timing	Oversight	Purpose	Observable Implications
Unilateral	Internal	Inconsistent and unpredictable	Political officials, military officials, and commanders from a *single* country	Country interests	• No consent or approval from other countries • Exclusive enforcement mechanism ("delegatory accountability")
Multilateral	External	Consistent and anticipated	Political officials, military officials, and commanders from *multiple* countries	International law and norms	• Consent and approval from other states • Inclusive coordination process ("participatory accountability")

based on a theory of victory against threats, particularly in undeclared or "cold" theaters of operations. This theory of victory relates to an assumption held by officials that strikes enable decapitation—the removal of charismatic leaders to hasten an enemy's demise—that is an effective way to achieve overall wartime aims while inuring their soldiers to battlefield harm (Hardy & Lushenko 2012; Pape 1996). Enhanced force protection helps minimize the reputational costs that officials may incur for using force abroad. A related intent of officials' strategic use of drones can be the restoration of countries' sovereignty. Indeed, drones often provide officials with the least bad option to provide for the internal security of fragile or failing countries. Though this threatens "strategic monism," which Huntington (1981) defined as a country's use of a single lever of its national power to achieve certain policy objectives, others are more sanguine about the prospects of the strategic use of drones. For these optimists, the strategic use of drones promises a cooperative way to achieve the common goal of security, but only if requested by targeted countries (e.g., Maurer 2022; Brunstetter 2021; Cannon 2020).

In practical terms, the strategic use of drones links limited resources—analysts, maintainers, crews and pilots, munitions, and drones themselves—to an operational approach—strikes—to achieve long-term outcomes, including decapitating an enemy or enhancing partnered military operations (Hazelton 2017). The observable implications of the strategic use of drones include the centralization of authority to conduct strikes within executive officials; the lack of reciprocal risk between combatants; a network of globally distributed bases to launch, recover, and maintain drones; and the potential erosion of sovereignty should intervening countries abuse the scope of their mission. This latter implication relates to an "imperial slide" that "occurs when protecting or restoring sovereignty slips into imperial drone use, thus marking the decision to override traditional sovereignty norms of other states" (Brunstetter & Férey 2022, 141). Whereas President Pervez Musharraf may have welcomed U.S. strikes in the Federally Administered Tribal Areas (FATA) of Pakistan against al-Qaeda in 2004, for instance, he reportedly withdrew consent shortly thereafter, although Obama expanded drone operations there (Ullah 2021; Benjamin 2012).

Constraint—Unilateral or Multilateral

Countries can also use drones with unilateral or multilateral constraints. Though both approaches may be designed to uphold the *jus in bello* principle of non-combatant immunity, they differ on the accountability of strikes for unintended consequences. Unilateral constraints are implemented by officials within a single country. This type of constraint, then, does not require the approval of other countries and is best characterized by "delegatory" accountability since countries assume sole responsibility to manage their use of drone strikes abroad. Indeed, Hodges conceives of this type of constraint as "military legalism." "When the legitimacy of a U.S. conflict is contested," he explains, "policy-makers are likely to implement rule-based regimes of constraint on the use of force in an effort to re-capture legitimacy" (Hodges 2018, iii). Enforcement of these internal targeting constraints is the remit of political and military officials responsible for theaters of operations. Sanctions for unintended consequences such as civilian casualties are also post hoc, occurring after mistakes are made, although the threat of reputational costs meted out by member states of international society can shape policies designed to prevent errors (Grant & Keohane 2005).

One example is Obama's Presidential Policy Guidance, announced on May 23, 2013, which the Biden administration reintroduced as well (Savage 2022). The policy emerged amid heightened criticism given the considerable harm imposed on civilians following U.S. strikes in undeclared theaters of operations, particularly Pakistan. Obama was initially comfortable striking targets based on the appearance of terrorist activity, which some scholars attribute to his lack of executive experience, which encouraged a militarized foreign policy (Calin & Prins 2015). These so-called signature strikes were criticized for unnecessarily harming civilians. Rather than synthesizing multiple types of intelligence—human, imagery, and signals—to confirm the identity of a target and its placement within a terrorist network, this model of signature targeting relied on a pattern of behavior that was characterized as terrorist given certain undisclosed criteria. This means that officials used strikes against targets "based on their intelligence signatures—patterns of behavior that are detected through signals intercepts, human sources and aerial surveillance, and that indicate the presence of an important operative or a plot against US interests" (Miller 2012).

Because signature strikes were perceived as indiscriminate (Silverman 2019), Obama adopted a policy that conditioned strike approval on the "near" certainty of no civilian casualties. As we show in Chapter 4, Obama's stringent targeting protocol resulted in a dramatic reduction of civilian casualties in Pakistan. Specifically, the policy resulted in a reduction of approximately 13 civilian deaths per month to one or less, enhanced the precision of U.S. drone strikes to 95%, and averted nearly 300 civilian deaths. Ansari's field research in the FATA, in which she interviewed 116 residents exposed to the U.S. drone program, shows that Obama's policy adjustment enhanced Pakistani citizens' perceptions of legitimate strikes, with one respondent claiming "[t]he drone is a justice-delivering technology" (2022b). In this case, Pakistani citizens seem to believe that U.S. drone strikes were legitimate in both descriptive and normative terms. This finding is corroborated by Shah's (2018, 58) trail-blazing field research in the FATA, showing that Pakistani citizens believe that "[d]rones are precise in most cases," as well as Mahmood and Jetter's (2023) use of a quasi-experimental design that uses wind gusts to explore Pakistani citizens' generally favorable sentiment for U.S. drone strikes.

Multilateral constraints obligate states to meet the oversight requirements of allies and partners given the purpose of international approval, which is to enforce laws and norms to legitimize operations (Kreps 2011). Strikes conducted under multilateral constraints typically demonstrate a shared responsibility across countries to ensure non-combatant immunity, though this does not mean that such measures will be effective in protecting civilians from battlefield harm. Multilateral constraints best relate to a "participatory" model of accountability where "the performance of power-wielders is evaluated by those who are affected by their actions" (Grant & Keohane 2005, 31). As we discuss in Chapter 5, the most well-known multilateral constraint is UN approval for countries' use of drones abroad. This is thought to impose stricter targeting protocols enforced during an inclusive coordination process involving political and military officials from many countries. This results in a negotiated process, usually under the aegis of a regional or international coalition of cooperating countries, for the use of strikes based on their anticipated advantages (Haun 2022).[7] Chief among these benefits is surgically removing terrorists while protecting friendly forces and preventing civilian casualties.

Cross-National Models of Drone Warfare

To be clear, there may be more uses of drones that are constrained differently (e.g., Rogers & Kunertova 2022; Rossiter & Cannon 2022). Our understanding of drone use and constraint reflects how most countries pattern their employment and management of strikes, as numerous other scholars also recognize (e.g., Chapa 2022; Phelps 2021; Boyle 2020; Cook 2015; Benjamin 2012). Even so, the tactical and strategic uses of strikes are not mutually exclusive. Of course, an individual strike does reflect a unique tactical or strategic purpose. Yet countries can integrate the use of tactically and strategically oriented strikes at the same time, in either the same or different theaters of operations. Similarly, countries can layer unilateral and multilateral constraints, as often happens during coalition military operations.

Disentangling how countries use strikes and under what constraints is partly a methodological choice, although it seems imperative. It would be challenging, if not impossible, to determine the causal relationship between countries' different uses of strikes under different constraints simultaneously, whether in the same or different theaters of operations, for the public's perceptions of legitimacy. This would be more problematic if multiple countries used drones within another country, as is taking place in Somalia where the United States, and now Turkey, are conducting complementary strikes against the al-Shabaab terrorist group that is affiliated with al-Qaeda (Caato 2022). The intent of our parsimonious framework, then, is to salvage a degree of complexity while also allowing us to understand how predominant patterns of drone use and constraint may moderate the public's perceptions of legitimacy in a cross-national context.

Variation in the use and constraint of drones suggests four patterns of strikes that we posit can shape public perceptions of legitimacy in predictable ways and cross-nationally (Figure 2.1). First, countries can use drones strategically with unilateral constraints. This "over-the-horizon" pattern of strikes implies countries use drones to attack terrorists without deploying boots on the ground. U.S. Army General (retired) Stanley McChrystal warns that this pattern of strikes creates the perception that "'we can fly where we want, we can shoot where we want, because we can'" (Alexander 2013). Indeed, this model best characterizes U.S. strikes, especially in undeclared theaters of operations, which Munro (2015), Gusterson (2015), Cachelin (2022), Enemark (2023), and many others refer to as a form of neo-colonialism. Since 2002, Brunstetter (2021, 47) notes that "the US self-defense-oriented drone paradigm, has, because of the lack of a shared norm governing this type of force, sometimes diminished cooperation." Yet U.S. officials have been more or less transparent about the protocols governing strikes and their effects.

Whereas Obama centralized approval for strikes within the White House, his successor, Trump, cashiered this stringent process after assuming office in 2017. His Principles, Standards, and Procedures delegated accountability to intelligence agencies and the military, which further clouded the requirements governing U.S. strikes and the outcomes (Regan 2022). Trump's strike against Iranian Major General Qasem Soleimani in January 2020 in Iraq, a veritable political official, evidences this rapacious model of strikes. So does Biden's botched strike in Afghanistan in August 2021. Instead of killing a suspected Islamic State terrorist amid America's withdrawal, the strike killed ten civilians, including women and children. To his credit, Biden's use of an over-the-horizon strike a year later in Afghanistan killed al-Qaeda's senior leader, Ayman al-Zawahiri, while resulting in no civilian casualties. This suggests that while Biden may favor strategic strikes with unilateral constraint, he does so while attempting to more scrupulously abide by International Humanitarian Law, given the putative legitimacy doing so affords, especially because it helps to mitigate the political fallout of preventable targeting errors (Engelbrecht & Ward 2022).

Second, countries' use of drones strategically with multilateral constraint constitutes "aerial occupation" (Emery & Brunstetter 2015; Benjamin 2012; Dudziak 2012). This pattern of strikes enshrines drones above other countries' geographically bounded territory, threatening to undermine their sovereignty. It capitalizes on the persistence and reusable nature of drones, which helps differentiate these capabilities from loitering munitions or "kamikaze" drones. This model of strikes is predominately practiced by the United States, but other countries, such as Turkey, have also increasingly exercised this approach (Hansen 2023).

Many scholars caution that aerial occupation is a euphemism for the respatialization of war that favors technologically superior combatants and is therefore unjust and racist. According to Gusterson (2014, 199), drones "scramble relations of distance, making them simultaneously more elongated and more compressed in ways that are subjectively confusing and paradoxical." Feldman also cautions that drones can constitute "racialization from above" because the targets tend to be Brown and Black people. The assertion that U.S. drone strikes are racially biased is liable to selection bias, however. The intended targets are predominantly Islamic terrorists in war-torn countries across Africa, Central and South Asia, and the Middle East that have darker skin. It is therefore challenging to assess the degree to which U.S. drone strikes may be racially biased without tapping into public attitudes, which is a task taken up by emerging research (e.g., Lushenko et al. 2023; Blakeley 2021; Feldman 2011). International authorization through the UN also means that the intent of aerial occupation shifts due to the strategic context of conflict, which can range from counterterrorism to humanitarian intervention, such as Obama's use of drones in Libya in 2011 to help arrest human rights violations by the Mu'ammar Al-Qadhdhāfī regime (Kreps & Maxey 2021). Schinella (2019, 260) finds that the drones "provided invaluable reconnaissance and strike capabilities" for that purpose,[8] which then Director of the Central Intelligence Agency, Leon Panetta, confirmed (Baldor 2011). The record indicates that U.S. drones conducted 145 strikes against enemy combatants and infrastructure, all with minimal civilian casualties (Rinehart 2018).

Sometimes, countries' use of drones can "fly under the radar" because strikes are conducted within their own borders and escape public attention (Kreps 2014). In other cases, countries use drones during border disputes. This tactical use of drones under unilateral constraint constitutes a "predatory" pattern of strikes because it threatens to militarize what should otherwise consist of law enforcement responses to political violence. This third pattern of strikes also threatens to escalate tensions between countries by providing a convenient *casus belli*. On the one hand, the National Transition Council of Libya used drones provided by Turkey—specifically, the TB2 Bayraktar—to strike rebels contesting the interim government following the U.S.-led coalition's withdrawal. One UN official even called Libya "the largest drone war theater now in the world" (Salamé 2019). On the other hand, Turkey also supplied TB2 drones to Azerbaijan during its dispute with Armenia in the contested Nagorno-Karabakh region. Though some analysts have called the TB2 a "game-changer" in the context of this conflict, others disagree, arguing that the existing power differential between the two countries was more influential for the negotiated settlement than drones (Calcara et al. 2022c). Nevertheless, analysts also claim that the TB2, now referred to as the "Toyota Corolla" of drones, is partly responsible for enabling Ukraine to shift the offense–defense balance during its war with Russia and has made Turkey a "drone superpower" (Soyaltin-Colella & Demiryol 2023; Khan 2022; Pitel & Jalabi 2022; Witt 2022).

Finally, countries that use strikes tactically with multilateral constraint, which presupposes international approval through the UN and almost always coalitions, adopt a "juridical" pattern of strikes. Some analysts, particularly Vilmer, refer to this as the "French model" of strikes. French officials use

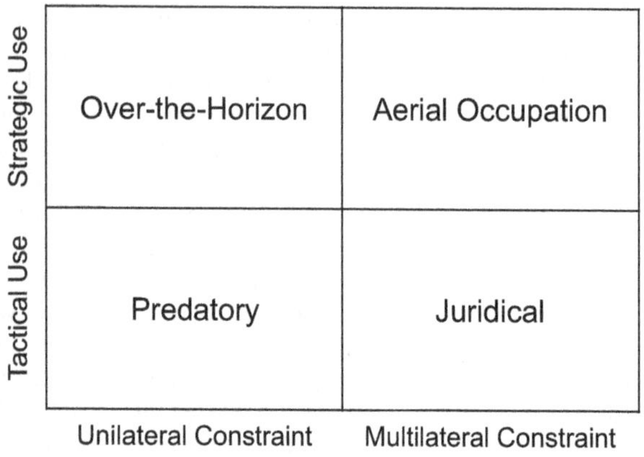

Strategic Use	Over-the-Horizon	Aerial Occupation
Tactical Use	Predatory	Juridical
	Unilateral Constraint	Multilateral Constraint

Figure 2.1 Cross-National Models of Drone Warfare

drones against terrorists in Western Africa under the sponsorship of regional and international organizations (Hamming 2023). This pattern of strikes is of special interest since it characterizes how a great power other than the United States uses drones beyond its immediate borders and region. Though scholars advance a French model of strikes, they do not investigate its genesis and the implications for the public's perceptions of legitimacy, which is a task we take up in the next chapter (e.g., Rogers & Goxho 2022; Vilmer 2021).

Conclusion

To summarize, drone warfare is often defined in terms of the platform, counterterrorism, or remote warfare. Yet countries often vary why and how they use strikes, meaning we can understand drone warfare as a function of unique and observable strike attributes, including variation in the use and constraint of strikes to help minimize civilian casualties. This definition suggests a typology of over-the-horizon, aerial occupation, predatory, and juridical patterns of drone warfare that allows us to explore the implications for the public's perceptions of legitimacy in a comparative context. In the following chapter, we use our novel middle-range theory to derive testable hypotheses for the public's perceptions of legitimate drone warfare among American and French citizens, considering their countries' leaders use drones prolifically across different regions of the world. In this case, we focus on two extreme models of drone warfare, over-the-horizon and juridical strikes, as an initial test of our middle-range theory, which is useful to advance the research agenda on public attitudes toward unmanned aerial systems. Indeed, we offer this analysis as an initial probe of the implications of evolving patterns of drone warfare globally, informed by our original middle-range theory, from which we encourage scholars to further study the legitimacy outcomes of other countries' varying use and constraints of strikes.

Notes

1 See also Elish (2017) and Kreps and Lushenko (2021).
2 See also Keating (2022).
3 Carvin also argues (2015, 128) that "the focus should be about the use of drones and similar weapons as a tactic—and to what extent they have evolved into a strategy."
4 Benjamin (2012, 108) also notes that drone warfare has been elevated to a veritable strategy as the best possible solution to the strategic challenges posed by non-state actors hiding in remote outposts of the world.
5 Though there are notable exceptions, however (e.g., Brooks & Wohlforth 2008; Carr 1949).
6 Though much more critical than Phelps (2021), Benjamin (2012, 18) also adds that drones mostly "patrol the skies looking for suspicious activity and, if they find it, they attack."
7 Haun (2022, 12) further explains that "[c]oalitions may deal with collateral damage concerns either by granting nations veto power over strikes or by allowing nations to opt out of particular missions. Nations may further place restrictions on the types of weapons their aircraft carry, such as when certain NATO [North Atlantic Treaty Organization] countries would not allow their aircraft to drop cluster bomb units (CBU) in Kosovo."
8 See also Matisek (2022, 177–201).

3 The Legitimacy of Drone Warfare in Comparative Context

The purpose of this chapter is to use our middle-range theory for evolving patterns of drone warfare, which we advanced in the previous chapter, to empirically test an expectation relating to the public's perceptions of legitimate strikes in a comparative context. Despite the proliferation of drones, scholars often interpret drone warfare as only a U.S. phenomenon. Yet other countries are acquiring drones and using them differently. Previous studies on public attitudes toward U.S. strikes, then, may not relate to other countries' use of drone strikes. Given this observation, we expect that citizens prefer distinct models of drone warfare that differ according to how countries use and constrain strikes to help protect against civilian harm. We focus on citizens from two countries that frequently use drones for counterterrorism beyond their borders and regions: France and the United States. We hypothesize that Americans perceive over-the-horizon strikes, which are characterized by the strategic use of drones with unilateral constraint, as the most legitimate. Similarly, we hypothesize that French citizens perceive juridical strikes, or those conducted tactically with multilateral constraint, as the most legitimate.

In the remainder of this chapter, we first justify our case selection and further elucidate juridical strikes. Though this pattern of drone warfare constitutes a clear counterpoint to U.S. over-the-horizon strikes, with France conducting several dozens of strikes against terrorists in the Western African country of Mali since 2019, scholars have yet to empirically test it in terms of public opinion. This includes the public's perceptions of legitimacy, which is the main outcome variable of interest we study. We then discuss our research design and methods, consisting of an original survey experiment and statistical analysis of the empirically derived data. Finally, we present our findings. The analysis of the data shows strong statistical support for our theoretical expectations. We also offer the first direct evidence of the French model of strikes as well as Americans' preferred pattern of drone warfare in terms of perceived legitimacy and identify instances where perceptions of legitimacy and attitudes of support deviate. This reflects what we call a "legitimacy paradox" and helps clarify that legitimacy matters in the context of public attitudes toward drone warfare.

DOI: 10.4324/9781032614267-3

Case Selection and the French Model

Since the terrorist attacks of 9/11, the United States has become synonymous with drone warfare. It is an incontrovertible fact that the United States has conducted more strikes and across more continents than any other country, even if the results are not widely known and officials have also dissembled publicly about the scope and scale of the U.S. drone program (Kreps et al. 2022). At the same time, other countries, some of which have purchased U.S.-manufactured armed and networked drones, namely the General Atomics MQ-1 Predator and MQ-9 Reaper, have also adopted drones as a principal counterterrorism tool. France is one of them, which makes it a useful comparison for several reasons. Since 2019, France has conducted numerous drone strikes in Mali. The most visible strike killed the Islamic State's leader in Western Africa, Adnan al-Sahrawi, in August 2021. He was the mastermind behind the deaths of French and Nigerian aid workers as well as four U.S. military personnel in 2017 (Maclean 2020). French attitudes toward strikes, then, also provide one litmus test for European preferences given the comparative decline in Britain's strikes abroad, Germany's hesitancy to arm drones, and Italy's limited operations. Finally, though scholars advance a French model of strikes, they do not empirically investigate it in terms of public opinion, including the public's perceptions of legitimacy. Doing so is important because it provides leverage over the public's mercurial treatment of strikes.

Public opinion for French and U.S. strikes in Africa reflects this puzzling trend. Despite being conducted in the same area, against the same threat, with the same drone, and with the same outcomes, including civilian casualties, public opinion varies dramatically in terms of French and U.S. strikes in Africa. Controlling for France's use of drones in Africa starting in December 2019, a simple search via *LexisNexus* reflects that French strikes receive nearly 60% less coverage than U.S. strikes. Whereas French strikes are generally celebrated by popular media outlets such as the *New York Times*, the United States' strikes are largely chastised (Maclean 2020). The accounting for French and U.S. strikes in Africa is perhaps more revealing. Though multiple watchdog organizations such as the Bureau of Investigative Journalism and New America exist to investigate U.S. strikes, no such organizations tally the locations, frequency, and effects of French strikes. The same is true of journalists, including those like Azmat Khan, who have been celebrated for investigating the dubious effects of U.S. counterterrorism strikes but at the expense of a more catholic treatment of drone warfare (The Pulitzer Prizes 2023). While some scholars relate the public's differential treatment of French and U.S. strikes in Africa to a French model of drone warfare, which we refer to as juridical strikes, it is unclear what this means, and the anecdotal evidence is difficult to substantiate (Vilmer 2017). Next, we further clarify the French model of drone warfare before turning to an empirical test of the public's perceptions of its legitimacy given the evolving patterns of drone warfare globally.

Though France purchased MQ-9 Reaper drones from the United States in 2013, officials in Paris were initially reticent to arm them. They sought to avoid the type of animus that the United States endured globally for collateral damage while also capitalizing on the intended dividends of drones. Eventually, France did start to arm its drones in 2017. Technical requirements delayed the retrofit process for two years, however. By 2019, and only after receiving a formal request from Mali, France started using drones against terrorists in Western Africa (Hollande 2013). Since then, France has conducted dozens of strikes against al-Qaeda and Islamic State terrorists, all with the endorsement of UN Security Council resolutions as well as regional partners.

France's decision to weaponize drones, then, was a matter of political will rather than technological wherewithal (Gilli & Gilli 2016; Horowitz et al. 2016). The justification was cast as a special managerial—if not post-colonial—responsibility (Holeindre 2018). Among other questions, French officials asked: "If the current situation of French-American relations permits us to accept, today, a certain degree of dependence, can we, in the short to long term, be permitted to consent to this enduringly?" French defense experts further contended that "it seems necessary and urgent to win the gamble on European drones . . . which is the utmost importance for European defense" as well as France's interests in Francophone Africa (Perrin et al. 2022). France's Defense Minister, Florence Parly, reasoned France's "methods and equipment must adapt" to contend with a "more furtive, more mobile" enemy that "disappears into the vast Sahel desert and dissimulates himself amidst the civilian population" (Irish 2017). To manage the potential for blowback, France has adopted a unique model of strikes.

For France, drones "constitute merely a tactic as opposed to a strategy they have come to represent for the U.S." (Brunstetter & Férey 2022, 146). Vilmer (2016, 7) adds that

> France disapproves the presumption that a state is always more likely to accept risk with a drone than with a human-inhabited aircraft, because it depends on external factors such as cost, the number of available units, and the strategic and political context.

In Mali, France often tasks its drones to gather intelligence on terrorists that enables strikes by fighter jets, ground-based raids, and operations by partnered forces (Dewan 2021). On March 3, 2022, for instance, the French military used intelligence provided by drones to conduct a raid in Mali that killed Yahia Djouadi, an al-Qaeda official (France 24 2022). When France does use drones against terrorists, it conducts strikes with multilateral constraint. One French defense official argues "[w]e need this multinationality to legitimate our actions" (Bentégeat, as cited in Schmitt 2018, 13). In Mali, France uses drones under the Multidimensional Integrated Stabilization Mission (MINUSMA), initially authorized by UN Resolution 2085 adopted in

December 2012. This coalition was expanded under UN Resolution 2100, adopted in April 2013, to help synchronize military operations by the "Sahel 5"—Burkina Faso, Chad, Mali, Mauritania, and Niger—and France's Operation Barkhane. In June 2021, MINUSMA was reauthorized by UN Resolution 2584. This further deputized France "to use all necessary means," including drones, to target terrorists while protecting against civilian casualties. By August 2022, France's president, Emmanuel Macron, withdrew French troops from Mali due to political tensions with the Malian government, repositioning them in the neighboring country of Niger. There, French forces transitioned from Operation Barkhane to establish the Takuba Task Force (King 2023).

French scholars celebrate the tactical use of strikes with multilateral constraint as a French model that is "much more harmonious, in terms of respect for humanist values" (Vilmer 2021). Rogers and Goxho (2022, 15) add that "French policy makers prefer to deploy a multilateral, reduced and remote capability to the Sahel, which gives a strong indication that remote light-footprint operations . . . are considered to be a more politically suitable type of intervention." What this means, according to Renic (2023b), is that the French approach to drone warfare better aligns the country's values with its interests or goals, consisting of, in this case, regional security and stability.[1]

For political theorists, this pattern of strikes implicates normative and instrumental trends that have shaped France's interventions abroad since the Cold War. The French Revolution (1789–99), which helped instantiate the norm of popular sovereignty, has encouraged French leaders to identify the country as a bastion of human rights (Bukovansky 2002). This has formed a link between France's obsession with prestige (*rang*) and humanitarian interventions. France attempts to reconcile these potentially incompatible goals by adhering to multilateralism, which was officially adopted as a foreign policy principle in 1994 (Recchia 2020; Staunton 2020). Kagan (2022) further contextualizes the French model against a

> European interest in inhabiting a world where strength doesn't matter, where international law and international institutions predominate, where unilateral action by powerful nations is forbidden, where all nations regardless of their strength have equal rights and are equally protected by commonly agreed upon international rules.

We advance the literature on France's approach to drone warfare by testing it in terms of perceived legitimacy and compared to U.S. preferences, which we also posit relate to over-the-horizon strikes.

Data and Experimental Design

Our research design consists of a 2×2×2 factorial and between-subject survey experiment. This includes eight treatment groups and a control group. The treatment groups vary three conditions across two axes: (1) use of a strike (tactical *or* strategic), (2) constraint of a strike (unilateral *or* multilateral), and (3) the unintended consequence of a strike (civilian casualty *or* not). We administered the survey among representative samples in France and the United States between November 2 and 16, 2021, via Qualtrics. The respondent pool totaled 1,823 and drew on 914 American and 909 French citizens. The full survey scenario, questionnaire, and summary statistics are included in Appendix A.

Our research design is advantageous for several reasons. First, it reflects how people make judgments in the real world, meaning independently or for one event at a time along a broader game tree of decision points (Kneer & Machery 2019; Bartels et al. 2015; Tversky & Kahneman 1981). By presenting respondents with one randomized vignette rather than multiple prompts, as is the case with choice-based survey designs, we reduce their cognitive load and isolate the implications of shifting strike attributes on perceptions of legitimacy. We also mitigate the possibility of measurement error on the basis of intra-respondent reliability (Clayton et al. 2023). Second, using Qualtrics to source representative respondent pools further reduces the bias associated with other online recruitment protocols. Randomized controlled trials using online convenience samples are subject to selection bias, considering the respondents are generally not reflective of the target population (Ternovski & Orr 2022; Berinsky et al. 2012). To avoid such endogeneity, we block respondents based on age, education, and gender, given census data gathered in France and the United States. Third, randomly and evenly distributing respondents across the experimental groups resolves the need to include control variables to draw inferences about the implications of varying strike attributes on perceptions of legitimacy. Finally, we incorporate feedback from a pilot of the survey to help refine the instrument.

While surveys are useful for isolating the effect of an experimental manipulation on some outcome of interest, they are sometimes criticized for resulting in biased findings, especially in cross-national contexts. Verba (1971, 310) cautions that the results may be "invalid because the measurements in the two societal contexts are not comparable." Further, the results can be challenged by response instability and treatment effects resulting from the scenarios, word choice, and question order. Together, these challenges amount to framing effects, often referred to as priming and social desirability bias, that can distort intuitions (Mutz 2022; Sinnott-Armstrong 2008). In this case, respondents are encouraged to answer in a certain way or do so because they feel obligated. At the same time, analyses of "cross-cutting"—or within-subject—surveys, such as conjoint-based designs, are often criticized for discounting the implications of interacting experimental groups on the overall treatment effects (Muralidharan et al. 2023).

We managed these concerns in several ways. First, we hired one translation service to translate the survey from English into French and then contracted another service to back-translate the instrument from French to English, which ensured the equivalency of the scenarios and questions. Second, we pretested the survey instrument to ensure it incorporated precisely worded and unambiguous questions, encouraging respondents to adopt the same frame of reference. Third, we adopted an "abstract encouragement" strategy to write the survey scenario (Chong & Druckman 2007). We wrote the scenario in terms of a hypothetical but realistic drone strike conducted by a fictional country, Country X (Dafoe et al. 2018). This technique is useful to "maintain the contextual grounding of the measures when making comparisons" (Verba 1971, 314) across respondents drawn from different countries. Research also suggests that the use of fictitious country names does not impose bias that can confound the results. Rather, this approach can actually increase the validity of the results by reducing the potential for priming (Majnemer & Meibauer 2023; Suong et al. 2023; Brutger et al. 2022). Fourth, we specifically do not interact with experiment groups in our statistical analysis given our between-subject design, meaning respondents get only one prompt, and we are interested in the main effects of these treatments on perceptions of legitimacy. Indeed, we have no intuition that different treatments, which are orthogonal to each other, somehow work together to shape perceptions of legitimacy. Finally, similar to the existing survey research, we use our vignette to provide brief definitions of the use and constraint attributes, which are consistent across all randomized experimental groups and therefore protect against the potential for priming (Tomz & Weeks 2020).[2]

Following the vignette, we probe the dependent variable. We ask respondents to rate their perception of how legitimate a strike was based on variation in use, constraint, and unintended consequence. We ask, "on a scale of 1 to 10, with 1 representing '**not** legitimate' and 10 representing '**very** legitimate,' how legitimate is Country X's use of the drone strike?"[3] At the same time, we attempt to gain leverage over the microfoundations that may mediate citizens' understanding of legitimate drone warfare. We gauge respondents' preferences for the use of force and support for great powers, as well as their beliefs that a strike should be domestically authorized, comport with international law, and minimize civilian casualties. This analysis builds on research for public support of drones in both France and the United States (Fisk et al. 2019).

Experimental Results

Overall, the results offer strong support for our hypotheses. Variation in countries' use of a strike tactically or strategically, coupled with how a strike is unilaterally or multilaterally constrained to protect against unintended consequences, moderates American and French perceptions of legitimacy in ways that we hypothesized that they would (Figure 3.1). When aggregating across all treatment groups, we use an analysis of variance to show that the relationship between the strike attributes of use, constraint, and unintended consequences for legitimacy outcomes is statistically significant for both American ($p = 0.000$) and French ($p = 0.001$) respondents. The extent to which respondents perceive strikes as legitimate is also remarkably consistent cross-nationally. Respondents perceive strikes as legitimate, rating their perceptions at nearly 7 out of 10 and with virtually the same standard deviation of 2.45.

However, clear differences emerge when analyzing respondents' perceptions of a legitimate strike by treatment groups. These results are important as they empirically validate, for the first time in the drone warfare scholarship, distinct patterns of strikes that respondents perceive differently in terms of legitimacy. The results also provide the first direct evidence of a purported French model of strikes as well as Americans' preferred type of over-the-horizon strikes. We also find that while legitimacy and support may strongly correlate, as public opinion researchers sometimes assume, these two attitudes often do not. This indicates a "legitimacy paradox," where respondents' perceptions of legitimacy and attitudes of support do not align. Below, we discuss these findings in greater detail. First, we place respondents' preferred model of strikes in a cross-national context. Second, we expose a "legitimacy paradox" with respect to respondents' preferred patterns of strikes. Third, we use statistical methods to disentangle, validate, and explain the French model in an effort to contribute new knowledge about this emerging pattern of drone warfare that represents the starkest contrast to U.S. strikes abroad, at least currently.

Cross-National Preferences for Strikes

When responding to separate questions, both sets of respondents endorse the tactical use of strikes as well as multilateral constraint, and at statistically significant levels based on one-tailed T-Tests ($p = 0.000$), which are designed to determine if the mean outcomes of groups of data are statistically different from one another. When aggregating across all treatment groups, a majority of American (68.5%) and French (59.3%) respondents also endorse non-combatant immunity as a key attribute for a strike. Multivariate regression analysis presented in Appendix A shows that this results in over a half-point increase in the perceived legitimacy of a strike in both America ($\beta = 0.68$, $p = 0.01$) and France ($\beta = 0.55$, $p = 0.01$). At the same time, Figure 3.1 reflects that both American and French respondents discount the strategic use of a strike with unilateral constraint that results in a civilian casualty (treatment three), and at the same level of statistical significance ($p < 0.05$). Notwithstanding these similarities, the results suggest that variation in strike attributes does moderate the public's perceptions of legitimacy in predictable ways.

First, American respondents mostly prefer the over-the-horizon pattern of drone warfare (treatment four) characterized by the strategic use of strikes with unilateral constraints that result in no civilian casualties. This finding is based on a T-Test (two-tailed) that estimates the difference in mean legitimacy

(a)

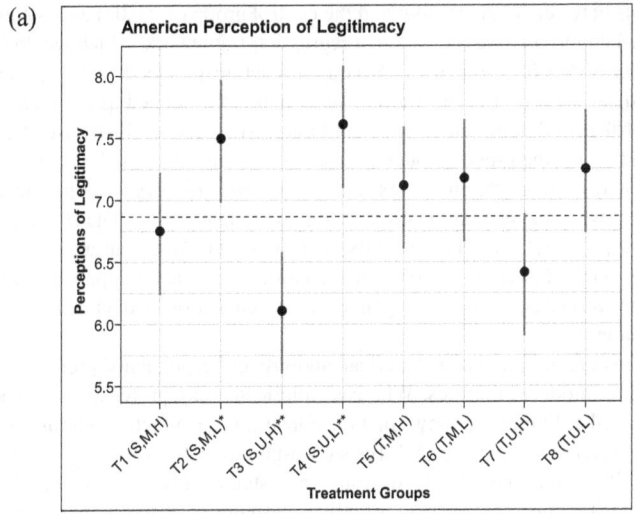

(b)

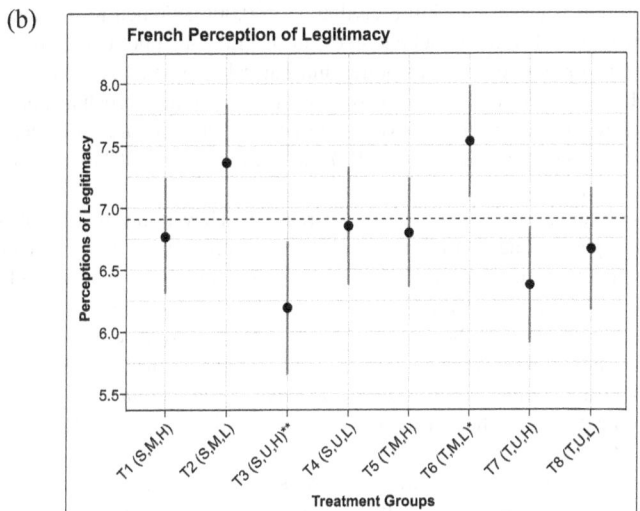

Figure 3.1 Respondents' Perceptions of Legitimacy by Treatment Groups

Note: Graphs depict American and French perceptions of legitimacy given different models of drone warfare while using the control group as a baseline (the dashed line represents the control group's mean legitimacy score as a reference point). Vertical I-bars present 95% confidence intervals about each legitimacy outcome. Statistical significance was calculated using Welch's T-Test (two-tailed) that estimates the difference in mean responses for treatment groups relative to the control. "S" refers to "strategic." "T" refers to "tactical." "M" refers to "multilateral." "U" refers to "unilateral." "H" refers to a "high"—or one—civilian casualty outcome. "L" refers to a "low"—or no—civilian casualty outcome. Statistical significance is represented by *** $p < 0.01$, ** $p < 0.05$, and * $p < 0.1$.

outcomes between treatment four and the control group ($p = 0.02$) and is consistent when interpreting the data in a regression framework, which we do in Appendix A ($\beta = 0.52$, $p = 0.10$). Second, French respondents mostly prefer the juridical pattern of drone warfare (treatment six), defined as the tactical use of strikes with multilateral constraint that result in no civilian casualties. This finding for the French model of drone warfare is also based on a T-Test (two-tailed) between treatment six and the control group ($p = 0.06$) and is consistent when using a multivariate regression to analyze the data ($\beta = 0.62$, $p = 0.10$), also reflected in Appendix A. In both the American and French contexts, our results are consistent when controlling for demographic and dispositional variables, including respondents' age, education, gender, and political ideology.

The results further clarify American and French respondents' preferences for distinct models of strikes. Whereas Americans may prefer the strategic use of a strike, Figure 3.2 shows that the constraint is, in fact, conditional on the consequence. According to T-Tests (two-tailed), Americans' preferences for multilateral or unilateral constraint are statistically indistinguishable when a strike does not result in a civilian casualty ($p = 0.67$) and statistically different when it does kill a civilian ($p = 0.006$). French respondents do not make this distinction but endorse multilateral constraint unconditionally. That is, French respondents prefer multilateral constraint as a matter of course. Using T-Tests (two-tailed) here as well, we find that French respondents' preferences for multilateral versus unilateral constraint are statistically significant when a strike does ($p = 0.04$) and does not ($p = 0.004$) result in a civilian casualty.

These findings are important for two reasons. They imply that the categorical imperative against killing civilians may apply more to drones than it does to nuclear weapons, which Dill et al. (2022) find does not shape public attitudes for the use of force in a cross-national—France and the United States—context.[4] Indeed, Sewall (2016) notes that the promise of precision munitions, that of preventing civilian casualties in war, seems to have shifted the debate from civilian harms proportional to anticipated military advantages to the mitigation of civilian casualties altogether, which Moyn (2021) echoes in a recent book.

These findings also suggest that Americans adopt a probabilistic approach to judging the legitimacy of strikes, where the civilian casualty outcome shapes perceptions of legitimacy. Indeed, research by Rowling and Blauwkamp (2021, 14) suggests that Americans may discount unilateral constraint when strikes kill civilians because such collateral damage undermines "the US's record on human rights and its fidelity to the rule of law." French citizens seem to emphasize the foresight of officials to do everything possible to minimize civilian casualties before taking a strike, hence their endorsement of multilateral constraint as a matter of course. In determining the legitimacy of strikes, in other words, Americans demonstrate an overestimation bias, whereas French citizens reflect *ex post* reasoning in which the potential for

(a)

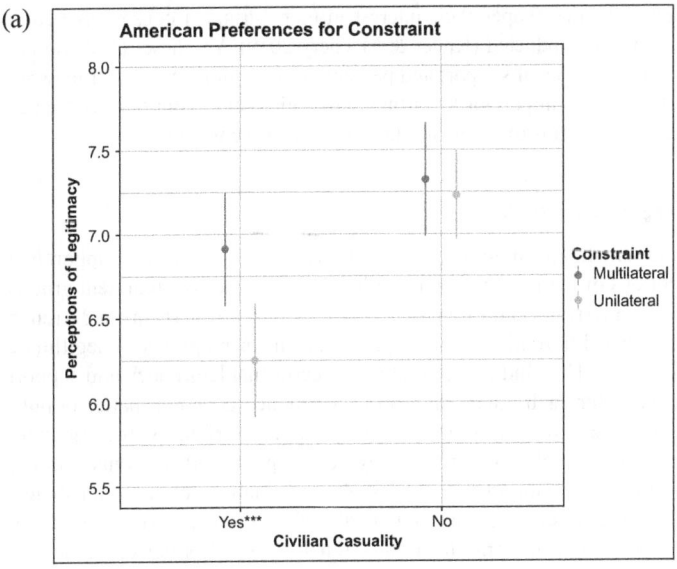

(b)

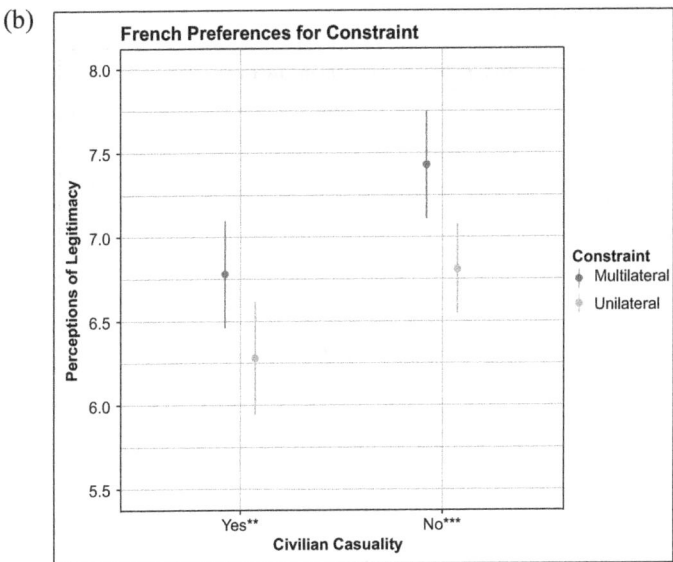

Figure 3.2 Respondents' Preferences for Constraint

Note: Graphs depict American and French perceptions of legitimacy given variation in the constraint and consequences strike attributes. Vertical I-bars present 95% confidence intervals about each legitimacy outcome. Statistical significance was calculated using Welch's T-Test (two-tailed) that estimates the difference in mean responses for each unique combination of strike attributes. Statistical significance is represented by *** $p < 0.01$, ** $p < 0.05$, and * $p < 0.1$.

civilian casualties shapes their interest in more stringent constraints before operations are conducted (Kneer & Machery 2019). Yet these results do not show how attitudes of support and perceptions of legitimacy may or may not covary, which is important to further substantiate our claim that legitimacy does matter in terms of public attitudes toward drone warfare.

The Legitimacy Paradox

We find statistically significant differences between attitudes of support and perceptions of legitimacy when pooling responses across treatment groups for both Americans and French citizens ($p < 0.01$). Americans and French citizens can differentiate their support for and perceptions of legitimate drone strikes. This finding helps allay concerns that legitimacy and support are proxies for each other, suggesting legitimacy does not matter in public opinion research and in the context of drone warfare. At the treatment group level, we observe greater degrees of perceived legitimacy rather than attitudes of support for all possible combinations of strike attributes, though the difference is only statistically significant in certain cases that shed further light on American and French citizens' preferred models of strikes (Figure 3.3).

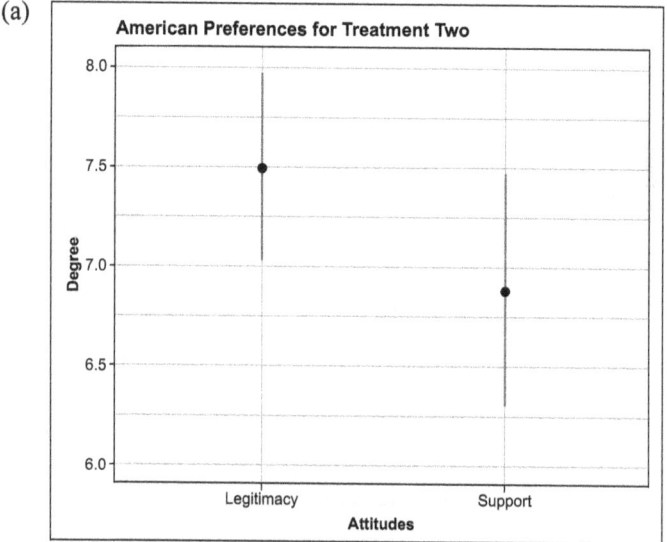

Figure 3.3 The Legitimacy Paradox

Note: Graphs depict American and French differences in perceptions of legitimacy and attitudes of support when varying the constraint and casualty attributes, respectively. Vertical I-bars present 95% confidence intervals about each legitimacy outcome. Statistical significance was calculated using Welch's T-Test (two-tailed) that estimates the difference in mean responses between the control group and treatment two for American respondents and treatment group five for French citizens.

(b)

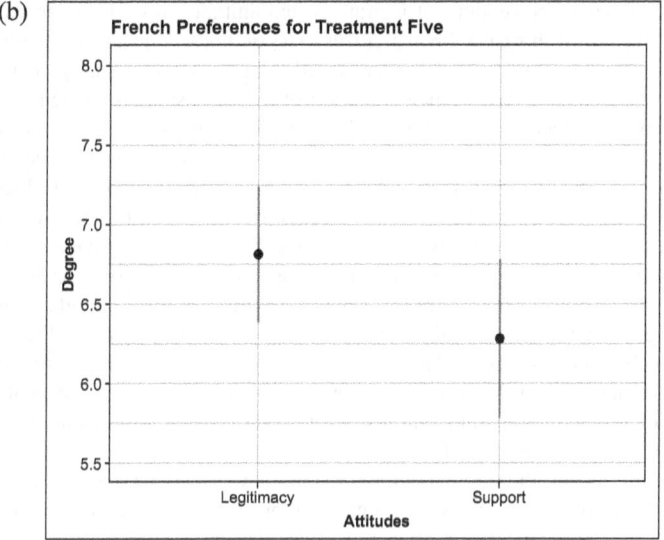

Figure 3.3 (Continued)

Americans perceive the over-the-horizon pattern of drone warfare—treatment four—as most legitimate and support this model of strikes the most as well. Indeed, the correlation between these two outcomes for treatment group four is positive and high at over 0.80, according to a simple—Pearson's—correlation test ($p < 0.001$). At the same time, Americans demonstrate important perceptual differences when strategic strikes are coupled with multilateral constraint, as they are in the aerial occupation pattern of drone warfare. They perceive this pattern of drone warfare, especially with the low casualty outcome represented in treatment two, as highly legitimate but support it less ($p < 0.10$). This suggests that while Americans favorably perceive the legitimacy of a pattern of drone warfare that other countries may equally endorse, as was the case with Obama's use of drones during the humanitarian intervention in Libya, this model of strikes ties U.S. officials' hands, reducing their freedom-of-maneuver globally and thus encouraging Americans to discount their support for such strikes. We further substantiate this finding using a logit regression model, presented in Appendix A. Specifically, we find that U.S. citizens are nearly 30% likely to reflect a "legitimacy paradox" for treatment two wherein respondents perceive this model of strikes as legitimate but do not support it.

French citizens' perceptions of legitimacy and attitudes of support also correlate highly in terms of their preferred model of juridical strikes, characterized by the tactical use of drones with multilateral constraint, especially when the civilian casualty outcome is low. The correlation between these two outcomes for treatment group six is positive and high at above 0.80, according to a simple—Pearson's—correlation test ($p < 0.001$). Similar to Americans,

French citizens' perceptions of legitimacy and attitudes toward support can also diverge. In this case, French citizens cue on non-combatant immunity, that is, civilian casualties. When presented with a civilian casualty, represented in treatment five, French citizens discount their support, though they may still view tactical strikes with multilateral constraint as highly legitimate ($p < 0.10$). To the extent that French perceptions of legitimacy and attitudes of support diverge, it appears that civilian casualties exercise an important effect. Similarly, we use a logit regression model to corroborate these results. We find that French citizens are nearly 30% likely to reflect a "legitimacy paradox" in terms of treatment five. Respondents perceive this model of strikes as legitimate but do not support it as much. Thus, across both samples and relative to their preferred model of strikes, the results show that American and French respondents can understand legitimacy and support differently, which further clarifies the relationship between these two outcomes of interest in public opinion research, at least among two great powers that are also democracies and often use drones abroad.

While useful, these findings do not show how American and especially French respondents actually interpret variation in the use, constraints, and unintended consequences of strikes. It is also unclear if French respondents can identify the French model when presented with a randomized treatment for this pattern of strikes that does not identify the targeting country. Further, the results do not indicate the microfoundations, or core values and beliefs, that may underlie respondents' preferred model of drone warfare. The latter consideration is important to help shed additional light on the "legitimacy paradox," which we observe among both American and French respondents. Below, we use a battery of statistical methods that allow us to triangulate the results to answer these important questions.

Explaining the French Model

To determine if French respondents identify the French model when confronted with a randomized treatment for the juridical pattern of drone warfare as well as the implications for perceptions of legitimacy, we use a combination of bivariate and logit regression models. A bivariate regression allows us to isolate the effect of French respondents' interpretation of the French model on their perceptions of legitimacy. We operationalized this model by first reviewing responses to the following open-ended question, which Verba (1971) notes is a helpful technique in cross-cultural survey research. "What country do you think Country X represents?" We constructed a binary variable, coding it 1 in the case French respondents identified Country X as France and 0 otherwise. Next, we regressed this binary variable on the legitimacy outcomes for treatment six, which we define in terms of a juridical model of drone warfare. The results suggest that French respondents not only identify the juridical pattern of strikes as the French model when given no indication of the targeting country, but doing so results in approximately a 1-point increase in their perception of legitimacy ($\beta = 0.96$, $p = 0.10$). Relative to the legitimacy outcomes depicted in the multivariate regression model (provided in Appendix A), this is also a substantively important finding, constituting 40% of the one standard deviation gain for the legitimacy score when aggregating across all treatment groups.

We also used logit regression models to isolate the implications of the French model for French respondents' perceptions of legitimacy.[5] We use this specification in four main ways, which is consistent with existing research on public attitudes about emerging technologies.[6] We regress each treatment group against French citizens' responses to the open-ended question for the country responsible for the strike to determine the probability that French respondents identify the country as France. Next, we regress strike attributes on French citizens' responses to the open-ended question for the country responsible for the strike to determine the probability that the constraint attribute is most associated with France. Given the French respondents' preferred model of strikes, we defined the constraint attribute in terms of multilateral constraints. We then regress French citizens' responses to the open-ended question for the country responsible for the strike on the legitimacy outcomes to determine what the projected country means for the probability that the strike is considered legitimate. We reverse this relationship in the final analysis, allowing us to determine how legitimacy fluctuates with country identification.

Together, the results empirically support the French model of strikes (see Appendix A for graphs of the statistical results). Further, they suggest an in-group and an out-group where strikes conducted by a respondent's own country are perceived as most legitimate. First, the probability that a French respondent identifies the striking country as France for treatment six—the French model—is approximately 25%. This is the highest-probability outcome of any treatment group. While this finding is not statistically significant at conventional levels, it is substantively revealing, considering it is consistent with the body of our findings for French perceptions of legitimacy for this unique pattern of drone strikes. Second, the probability that a French respondent identifies the striking country as France is 22% when the strike is multilaterally constrained and 17% otherwise ($p = 0.10$). Third, the probability a French respondent identifies the striking country as France is 22% when the strike is perceived as legitimate and 10% otherwise ($p = 0.005$). Finally, the probability that a French respondent identifies the strike as legitimate is 89% when the striking country is France and 76% otherwise. This is also highly statistically significant ($p = 0.005$). It is difficult to say for sure if these findings reflect certain preferences for French drone strikes abroad or merely the public's general knowledge of these operations. The sterilized—fictional—nature of the survey scenario does, however, help substantiate the claim that French citizens can empirically identify their preferred model of juridical strikes, which they associate mostly with France and perceive as more legitimate than other patterns.

Importantly, these results are similar but more pronounced among American respondents (see Appendix A). Americans also identify their preferred model of strikes, consisting of the strategic use of drones with unilateral constraints resulting in no civilian casualties, at a rate of 88%. This is the highest rate of any treatment group and is statistically significant ($p = 0.03$). These results suggest that the country conducting the strike does matter for perceptions of legitimacy. To further substantiate this finding, we ran a final logit regression to assess the degree of ethnocentrism associated with varying models of strikes, which reflects a "predisposition to reduce all of social life to in-groups and out-groups" (Kinder & Kam 2010, vii). Americans' preferred model of strikes is 73% likely to be associated with ethnocentrism, which is the highest score registered by U.S. respondents for all patterns of drone warfare ($p = 0.10$). This is not entirely surprising since Kinder and Kam (2010, 84) find that "ethnocentrism plays a major role in motivating American support for the war on terrorism," with drone strikes being a key instrument. French citizens' preferred model of strikes is 42% likely to be associated with ethnocentrism, the lowest score registered by French citizens for any pattern of drone warfare ($p = 0.001$). While this finding may belie France's comparatively limited use of drones, it could also suggest unique microfoundations that mediate French attitudes toward strikes.[7]

What helps explain French citizens' preference for a unique French model of strikes? Where other scholars have explored this question, attributing the juridical pattern of drone warfare to France's "respect for humanist values," it is unclear what this means (Vilmer 2021). To the extent this claim is helpful, it is also non-falsifiable, difficult to replicate, and challenging—if not impossible—to generalize (Findley et al. 2020). It provides us with little indication of the degree to which French officials' putative interest in preserving human rights abroad constitutes—or is constituted by—citizens' shared belief in humanitarian interventions. It also cannot account for a "legitimacy paradox" where perceptions of appropriateness do not align with attitudes of support. To gain leverage over the microfoundations that may mediate French respondents' perceptions of legitimate strikes when they cohere with the juridical model of drone warfare, we turn to causal mediation analysis. As we discussed in the previous chapter, this method shows the complete causal chain for the effect of an independent variable on a mediator and the effect of a mediator on the dependent variable (Imai et al. 2011). In other words, it shows "the mechanism by which an exposure (treatment) affects an outcome, in particular by studying intermediate variables that might be responsible for transmitting such an effect" (Miles 2022).[8]

Though we ensure to fulfill important assumptions to operationalize this method, namely randomizing the order of survey questions for respondents across the control and treatment groups (Chaudoin et al. 2021), causal mediation analysis is sometimes criticized for failing to account for confounding—omitted—variables, even in an experimental setting that researchers usually champion for resolving stochastic or random error. Miles (2022, 2) argues that confounders "cannot be eliminated even in a well-controlled randomized experiment," causing Simonsohn (2022) to argue that scholars typically "over-estimate the mediators." In recognition of these valid concerns, our use of causal mediation analysis follows other studies that attempt to adjudicate public attitudes for war, including France's use of drones. These studies adopt what Imai et al. (2010) refer to as the "sequential ignorability assumption," whereby possible pretreatment confounders and treatment assignment are assumed to be orthogonal to, or statistically independent from, the potential outcomes and mediators (Montgomery et al. 2018).

One of these studies focuses on the effects of anger and fear among French respondents to explain their support for strikes, finding uneven results (Fisk et al. 2019). Rather, we draw potential microfoundations from the literature on war and multilateralism, especially the approval by the UN Security Council for the use of force.[9] This may allow us to further account for the microfoundations of French respondents' perceptions of legitimate strikes, particularly given Simonsohn's (2022) criticism that causal mediation analysis risks having the mediator(s) capture "the confounded effect of all things that correlate with studying outside the experiment." Indeed, scholars have established that perceived legitimacy can reflect both normative and instrumental dimensions for countries' use of drone strikes, but stressed that this needs more study (Kreps & Wallace 2016).

We identify French respondents' perceptions of legitimacy as the dependent variable and the French model as the independent variable. To further focus our analysis, especially given French respondents' preference for tactical strikes, we restrict our primary specification to treatment groups that reflect this use attribute (below, we incorporate all treatment groups as a robustness check). We also identified five microfoundations cited in the literature, which we included in the survey: domestic authorization, international law, morality, merit, and preference for the use of force. First, we asked French respondents to assess how important it is for Country X's legislature to authorize the use of force. Second, we asked French respondents to assess how important it is for Country X to uphold international law during the strike. Third, we asked French respondents to assess how acceptable civilian casualties are in war. Fourth, we asked French respondents to assess the role of great powers in global politics. Finally, we asked French respondents to assess the role of force in international relations. For all of these questions, French respondents selected one of five outcomes ranging from "very unimportant" or "strongly disagree" (1) to "very important" or "strongly agree" (5).

The results in Figure 3.4 reflect that normative considerations outweigh compliance with the law, both domestically and internationally. About 20%

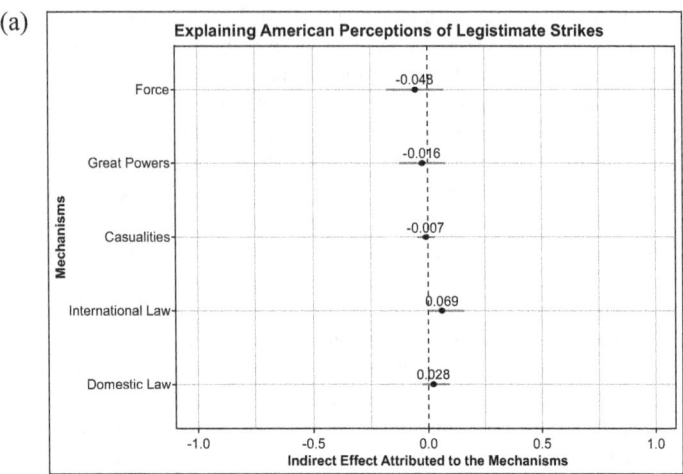

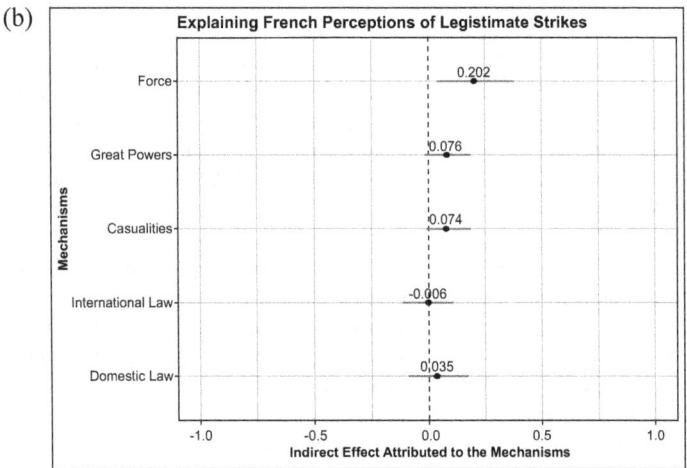

Figure 3.4 The Microfoundations of American and French Perceptions of Legitimacy

Note: Graphs depict microfoundations for American and French perceptions of legitimacy. Results were calculated using the "mediation" package in R. Horizontal bars present 95% confidence intervals about each point estimate. Bars that cross the dashed line at zero indicate no statistically significant results for the proportion of the treatment effect on the legitimacy outcome explained by the mediators. To calculate the proportion of the effect on the outcome explained by these mediators, we multiplied the effect of the treatment on the mediators by the effect of the mediators on the outcome and then divided by the total treatment effect. This approach replicates a method proposed by Baron and Kenney (1986) and used by others (Tomz and Weeks 2020; Fisk et al. 2019).

($p = 0.02$) of the indirect effect arose because French respondents reflect a preference for the use of force abroad, and another 7.6% ($p = 0.10$) arose because French respondents emphasize the special managerial role of great powers in global politics. Morality also constituted nearly 7.5% ($p = 0.06$) of the indirect effect, meaning that French respondents' perceptions of legitimate strikes are sensitive to civilian casualties. The mediating roles of domestic and international law are much weaker. While the result for international law is surprising considering multilateral constraint is subject to international approval through the UN, the broader finding seems to reflect that ideational considerations really do underpin the French model, which other scholars have claimed (see Appendix A for a table of the overall treatment effects). At the same time, it also appears that the French prefer to assume a post-colonial responsibility for previous colonies such as Mali, as others have also suggested (Holeindre 2018). Brunstetter (2022) relates this approach to a "drone contract" between France and other "decent" countries that are afforded equal status as rightful members of international society, which is itself a normative judgment.

Together, then, the normative considerations of morality, merit, and preference for the use of force abroad account for over a third of the indirect effect, whereas compliance with domestic and international law plays a smaller role in the causal chain. Comparatively, the results suggest that U.S. citizens emphasize their perceived compliance with international law, though the effect is marginal and more puzzling. Americans' preference for the strategic use of strikes with unilateral constraints that result in no civilian casualties often breaches other countries' sovereignty. This finding deserves more study, though it is consistent with previous research on Americans' ambivalence toward drone strikes (Kaag & Kreps 2014; Lushenko & Kreps 2022).

Though intuitive and useful, causal mediation analysis has no significance test. To assess the strength of the results, therefore, we ran two robustness checks. First, we used the Sobel-Goodman mediation tests. In effect, these are T-Tests that show whether the indirect effect of the independent variable—the French model—on the dependent variable—legitimacy—through the five mediators is statistically different from zero. This operation shows significant mediating effects for French respondents' preferences for the use of force ($p = 0.02$) and marginally significant mediating effects for their understanding of the morality ($p = 0.13$) and merit ($p = 0.14$) of a strike. Second, we broadened the analysis to include all treatment groups, which we initially restricted to those reflecting French respondents' preference for the tactical use strike attribute to further isolate the French model. In doing so, we find that French respondents' preferences for the use of force continue to account for a preponderance of the indirect effect (17%) and at a statistically significant level ($p = 0.03$). On the other hand, morality and merit further decline in effect and significance, amounting to 3% ($p = 0.27$) and 6% ($p = 0.16$) of the indirect effect, respectively. These results are not entirely surprising when considering all treatment groups, however. The strategic use of strikes is more consistent with an instrumental use of force that offsets normative considerations.

Implications

Our results suggest that the accepted practice of studying only U.S. strikes as a gauge of broader shifts in drone warfare globally, while understandable given data limitations, is misleading. It distorts the emergence of competing models of strikes across countries that belie unique strategic cultures that inform why and how countries use military force abroad. Data limitations have also encouraged scholars to probe Americans to track public attitudes toward strikes. This common practice imposes another trade-off, however. Notwithstanding the emergence of distinct models of strikes globally, which we show can be framed by varying use and constraints to help mitigate civilian harm, scholars have conflated U.S. attitudes with global public opinion. Moreover, attitudes are defined in terms of approval or support rather than perceptions of legitimacy, because people are thought to think more in terms of the former than the latter. This observation is ironic. As we discussed in the introductory chapter, political and military leaders justify their policies in terms of legitimacy; scholars define legitimacy as central to the sustainability of drone strikes; and research shows that support and legitimacy are not interchangeable, though both are social–relational concepts that help proxy for the public's beliefs in the use of force. Indeed, our results suggest the possibility of a "legitimacy paradox" in terms of French and U.S. preferences for certain models of drone warfare, whereby respondents' perceptions of strikes as more or less legitimate compared to attitudes of support.

Our findings are best thought of as an initial step to help scholars broaden the scope of their investigations into countries' adoption of drones. We contribute to the complicated discussion on the moral psychology of war by documenting how the publics of two countries that frequently use strikes for counterterrorism abroad can perceive shifting patterns of drone warfare as more or less legitimate. French respondents prioritize the legitimacy of a strike conducted tactically with multilateral constraints that safeguard civilians. We also find that French citizens prefer multilateral constraint, regardless of the anticipated civilian casualty toll from a strike. This finding is helpful to certify a wider body of scholarship that posits a unique model of strikes adopted by French authorities without ever empirically validating that this is actually the case. American respondents, on the other hand, prioritize the legitimacy of a strike conducted strategically, and the anticipation of civilian casualties moderates their preference for this type of constraint. An expectation for civilian casualties encourages Americans to emphasize multilateral constraint, whereas a "clean" strike resulting in no civilian casualties shapes their endorsement of unilateral constraint. These results suggest the need for more analysis on the constraint attribute in particular, which is a task we take up in the following chapters.

These findings have at least two important policy implications. First, leaders integrating drones into their military arsenals must be aware that the manner of their use and constraints can shape public attitudes toward strikes, including perceptions of legitimacy, which may have cascading effects in terms of domestic audience costs and international blowback (Shah 2018). A series of investigations by the *New York Times* into the harms of U.S. strikes abroad caused the Pentagon to develop the "Civilian Harm Mitigation and Response Action Plan," or CHMR-AP, which echoes the longstanding concerns of anti-drone advocates (Schmitt et al. 2022).[10] Whereas the French model of strikes may strive to reduce civilian casualties on account of multilateral constraints, it is also vulnerable to mistakes. Indeed, Silverman (2020) notes that "even the best-organized and best-trained armies will kill some civilians in any prolonged military engagement." A UN investigation into a French strike in Mali in January 2021, for instance, found that it killed at least 19 civilians. This caused the French Defense Minister, Parly, to reinforce France's commitment to multilateralism in the region (France 24 2021).

Second, aside from emphasizing one constraint over another, countries should deliberately integrate unilateral and multilateral constraints to prevent civilian casualties. U.S. officials can enhance the perceived legitimacy of their drone strikes by augmenting secretive negotiations with officials from intervening countries with approval from regional and global institutions, as is now happening in Somalia, which one observer has called "the center of a U.S. counterterrorism drone war" (Schmitt 2023). Officials from Turkey and the United States are increasingly cooperating with the African Union to use strikes, launched from both Turkish- and U.S.-manufactured drones, to disrupt al-Qaeda's command and control and freedom-of-maneuver (Hansen 2023). French officials can also adopt a near certainty standard of no civilian casualties during French strikes to protect against mistakes, which can happen in war even when the best precautions are taken. In the following two chapters, we further explore these constraints, focusing first on unilateral constraints (Chapter 4) and then multilateral constraints (Chapter 5). The purpose of our analyses is to provide an initial probe of these constraints in terms of achieving the intended effects of drones while also relating the outcomes to the public's perceptions of legitimacy.

Notes

1 See also Krasner (1978).
2 Our treatment is not "bundled," meaning we cannot be sure what shaped perceptions of legitimacy in our experimental setting because we varied more than the strike attributes.
3 We define this concept in the question prompt as "how right or wrong you perceive the strike to be."

4 Research also suggests that the military advantages of nuclear weapons—a decisive battlefield advantage, for instance—are dependent on the capability's military disadvantages, namely large-scale death and destruction (see Brown et al. 2023).

5 A logit model allows researchers to estimate parameters for a model with a dichotomous dependent variable ranging from 0 to 1 (Baily 2017).

6 See, for instance, Horowitz and Lin-Greenberg (2022) and Lin-Greenberg (2022).

7 We regress the treatment groups against respondents' answers to the following question, which we adopted from the World Values Survey (2017–21): "When jobs are scarce, employers should give priority to people of this country over immigrants." Responses ranged from "strongly disagree" (1) to "strongly agree" (5), which we dichotomized for a measure of ethnocentrism.

8 See also MacKinnon et al. (2007).

9 See, for instance, Lebow (2020), Tomz and Weeks (2019), Kreps (2010), and Voeten (2005).

10 See also Muhammedally and Mahanty (2022) and Lushenko (2022b).

4 Unilateral Constraint and Public Perceptions of Legitimacy

The findings from Chapter 3 suggest that Americans' perceptions of legitimate drone warfare can be moderated by the expectation of unintended consequences, an outcome that is shaped, in part, by officials' constraints on strikes. Indeed, the approval of military operations is subject to constraints intended to achieve a just war, a key principle of which is the protection of civilians from harm. Since World War II, International Humanitarian Law—or the Laws of Armed Conflict—has dictated that civilian harm may never be the objective of military action (Fazal 2018; Crawford 2003). Despite this prohibition, which is known as distinction or non-combatant immunity in legal parlance, the causal effect of constraint on civilian protection during U.S. drone strikes is poorly understood and studied. The lack of scholarly attention is exacerbated by America's preferred model of over-the-horizon strikes, which is generally characterized by poor transparency that altogether obscures the constraint attribute (Regan 2022).

In this chapter, we draw on our middle-range theory to further examine the implications of Americans' preferred model of over-the-horizon strikes. We exploit a dramatic shift in the unilateral constraint of U.S. drone strikes during the Obama administration and assess its effectiveness in terms of civilian protection in Pakistan while also incorporating existing research to determine what this may mean for the public's perceptions of legitimacy. Pakistan was the site of a preponderance of U.S. drone strikes during the Obama administration, making it a useful case to study in terms of the constraint attribute. Prior to the policy change, approval of U.S. drone strikes was constrained by a standard of reasonable certainty that no civilians would be killed, meaning associates of targeted terrorists had a strong presumption of *guilt*. We study whether the adoption of a different and more stringent unilateral constraint—the near certainty standard of no civilian casualties during U.S. drone strikes in Pakistan, reduced the incidence of unintended civilian harm. This change—formally announced by Obama on May 23, 2013 at the National Defense University in Washington, DC—was designed to enhance the legitimacy of U.S. drone operations in Pakistan and elsewhere (Lindsay 2020).

DOI: 10.4324/9781032614267-4

The remainder of this chapter unfolds into three parts. We first introduce our data. In doing so, we detail several rare interviews with senior Obama-era officials responsible for writing and implementing the policy, who confirm a two-year lag between the policy's implementation and formal announcement, which aligns with our statistical analysis. For the first time in the literature, we use a mixed-methods research design to provide a more accurate policy implementation date from which to gauge the true effectiveness of the near certainty standard and the implications for the public's perceptions of legitimacy. Indeed, previous assessments of the policy's effectiveness have distorted its timing, therefore precluding an accurate evaluation of this new targeting standard (Regan 2022; Sheehan et al. 2022; Boyle 2020; Benjamin 2012). Finally, we present descriptive evidence on the dramatic changes in strike outcomes before and after the Obama administration leveraged the near certainty standard. Rather than focus strictly on outcomes in terms of civilian casualties, we also discuss the social and economic implications of this more stringent unilateral constraint, which is based on our use of simulations. This is an important advancement in the drone warfare literature. Whereas scholars may also recognize the socio-economic costs of strikes (Richardson 2022; Slim 2022), most have not explored this dimension empirically, including the implications for the public's perceptions of legitimacy.

Data

We use data provided by the Bureau of Investigative Journalism (BIJ). This captures the universe of U.S. drone strikes in Pakistan from 2004 to 2018. The data is based on a triangulation of news reports, official statements, and government press releases. Specifically, the data captures 430 U.S. drone strikes and provides strike-specific information about their locations in Pakistan as well as their effects in terms of civilian casualties, child casualties, and total casualties.

We opt to use the BIJ dataset for several important reasons. First, the data provides multiple citations and external checks for each catalogued U.S. drone strike in Pakistan. The casualty values are validated by multiple sources, and when cross-referencing a subset of the data, we identified no reporting errors. Second, scholars, policy-makers, and practitioners often question the veracity of the U.S. government's estimates for civilian casualties during drone strikes abroad. The author of Obama's policy, for instance, admitted that "the U.S. figures undercount civilian casualties" (Hartig, personal communication November 5, 2021). Finally, other watchdog groups that also collect data on U.S. drone strikes, such as New America and The Long War Journal, are often criticized for inconsistencies. Kreps (2016, 22), for instance, notes that "civilian casualty figures have proven to be quite controversial, as there tends to be discrepancies across studies." Shah (2018, 66) also adds that "civilian casualty figures are also misreported or unreliable because they are based on information provided by unnamed Pakistani or U.S. officials who have an interest in exaggerating or underreporting the civilian death toll." The BIJ dataset, on the other hand, is regarded by many U.S. policy-makers as the most authoritative ledger of U.S. drone strikes in Pakistan (Mahmood & Jetter 2023; Hartig, pers. comm. November 5, 2021; Benjamin 2012).

Consistent with the prior literature, our outcome variables consist of the midpoint of the minimum and maximum estimates for reported casualties following U.S. drone strikes in Pakistan (Sheehan et al. 2022). We construct these midpoint values for civilian, child, and total casualties and also create a measure of combatant casualties as the difference between total and civilian casualties for any given strike. We also construct an outcome measure of strike precision from these values as the proportion of total deaths from a strike that are reported as combatant deaths. This measure of precision ranges from 0 to 1, taking the value of 1 when only combatants are killed and 0 when only civilians are killed. Strike precision, then, can be interpreted as the percent of total deaths from a strike that were combatants and provides further insight into the efficacy of the near certainty standard.

Policy Timing and Background

Under the Bush administration, U.S. drone strikes were conditioned on a lenient standard: reasonable certainty of no civilian casualties. As a result, civilian harm was incurred in nearly 40% of U.S. drone strikes, amounting to an estimated 630 civilian casualties. On May 23, 2013, the Obama administration publicly announced that it had transitioned to a stricter targeting standard for drone strikes in undeclared theaters of operations, including Pakistan. This policy—formally termed Presidential Policy Guidance—conditioned the approval of U.S. drone strikes on four requirements. A strike would be approved after demonstrating the following: (1) a target that constituted a "continuing imminent threat to U.S. persons"; (2) the infeasibility of capturing the intended target; (3) the near certainty of target identification; and (4) the near certainty of no civilian casualties. Obama's goal, according to senior officials we interviewed, was to encourage more precise strikes to enhance the perceived legitimacy of U.S. drone strikes abroad. By 2012, for instance, a Pew Research Center poll showed that 94% of Pakistanis thought that U.S. strikes in Pakistan were killing "too many" civilians (Fair & Hamza 2016). Obama's use of drones to kill terrorists, then, was understood as inappropriate by a majority of citizens in Pakistan, if not by the public in other countries across the region.

Validating the Implementation Date

We first validate the implementation date of July 2011 offered by senior Obama-era officials. These include the former Director of the Central Intelligence Agency, John Brennan; the former National Security Advisor, Thomas Donilon; and the former Senior Director of Counterterrorism at the National

Table 4.1 Descriptive Statistics for U.S. Drone Strikes in Pakistan

	Full BIJ Sample			Obama Administration		
	Certainty Standard			Certainty Standard		
	Reasonable Certainty	*Near Certainty*	*% Diff*	*Reasonable Certainty*	*Near Certainty*	*% Diff*
Panel A. Cumulative Strike Outcomes						
Strike count	276	154	−44.2	223	148	−33.63
Injuries	1,050	406	−61.33	820	403	−50.85
Deaths						
Civilian	630	66	−89.52	364	64	−82.42
Child	182	7	−96.15	60	7	−88.33
Non-civilian	1,726	848	−50.87	1,472	829	−43.68
Total	2,356	914	−61.21	1,836	894	−51.31
Panel B. Strike-Level Outcome Means						
Injuries	3.8	2.64	−30.53	3.68	2.72	−26.09
Deaths						
Civilian	2.28	0.43	−81.14	1.63	0.44	−73.01
Child	0.66	0.05	−92.42	0.27	0.05	−81.48
Non-civilian	6.25	5.51	−11.84	6.6	5.6	−15.15
Total	8.54	5.94	−30.44	8.23	6.04	−26.61
Panel C. Monthly Outcome Means						
Strike count	5.43	2.88	−46.07	7.67	3.13	−59.19
Precision	0.69	0.95	37.68	0.80	0.95	18.75
Injuries	20.12	7.63	−62.08	27.92	8.58	−69.27
Deaths						
Civilian	11.98	1.21	−89.90	12.27	1.33	−89.16
Child	3.46	0.12	−96.53	2.05	0.13	−93.66
Non-civilian	33.68	15.47	−54.07	51.03	17.11	−66.47
Total	45.66	16.68	−63.47	63.30	18.44	−70.87

Note: This table shows descriptive statistics for U.S. drone strikes in Pakistan, for both the full BIJ sample (2002–19) and the Obama administration (2009–17). We subset the data according to the reasonable and near-certainty standard that governed strike approval and identify cumulative, monthly, and strike-level outcomes for civilian harm, non-civilian harm, strike count, and strike precision. The civilian harm outcome measures both the total reported injuries and deaths for civilian adults and children. The non-civilian harm outcome measures the total reported deaths for non-civilian adults or combatants. The strike count measures the total reported frequency of strikes. Strike precision measures the proportion of total reported non-civilian or combatant deaths to total reported deaths, including children, following U.S. drone strikes in Pakistan. We calculate strike precision at the monthly level to provide a more meaningful estimate of the effect of the near certainty standard on reducing civilian harm defined in terms of civilian deaths.

Security Council, Luke Hartig (Lefante 2023; Brennan, personal communication March 14, 2022; Hartig, personal communication November 5, 2021). We identify a discrete change in the rate of civilian harm during U.S. drone strikes in Pakistan prior to May 2013. We do so by estimating local Poisson regressions with a 21-month bandwidth on either side of each month in our data. A Poisson specification is advantageous because it measures the probability of an outcome, in this case civilian casualties, happening in a discrete time period. We chose a 21-month bandwidth to capture observations that occurred after July 2011 but before May 2013 in an effort to identify the discrete change attributable to each date separately. We identify samples comprising the 21 months before and after each month in our data. Having done this, we iteratively estimate the following regression:

$$y_m = \exp\left(\alpha + \beta_1\, Month + \beta_2\, Month \times Post_{July2011} + \beta_3\, Month \times Post_{July2013}\right)$$

In this equation, y_m is the monthly average number of civilian casualties per strike. We model the average number of civilian casualties per strike to ascertain whether the changes are, in fact, driven by increased strike precision rather than fewer outlier strikes that killed many civilians. We interact our time variable *Month* with two binary indicators that take the value of 1 after July 2011 and May 2013 and 0 otherwise. We weight observations by the number of strikes conducted in a given month. Using each iterative subset of data, we predict the rate of civilian harm based on these local Poisson regressions. We observe a decreasing trend in predicted civilian harm prior to the July 2011 discontinuity, supporting existing explanations of reductions in civilian risk due to improvements in intelligence, technology, and targeting proficiency (Figure 4.1) (Plaw et al. 2011). Importantly, we also observe a discrete change in predicted values following July 2011. The first of these values is drawn from a local Poisson regression that does not include observations after May 2013. This evidence aligns with our interviews and validates, for the first time in the drone warfare literature, the policy's true implementation date of July 2011. This finding, again, is integral to determining the degree to which the near certainty standard achieved its intent as well as the implications for the public's perceptions of legitimacy.

Near Certainty and Strike Outcomes

Table 4.1 presents descriptive statistics for U.S. drone strikes approved under the reasonable and near certainty standards for the full BIJ sample and separately for strikes conducted during the Obama administration. Panel A shows that the United States conducted 430 drone strikes in Pakistan from June 2004 to January 2018, which injured 1,456 civilians and killed 696 more. Panel B reveals that, on average, a strike conducted under a reasonable certainty standard

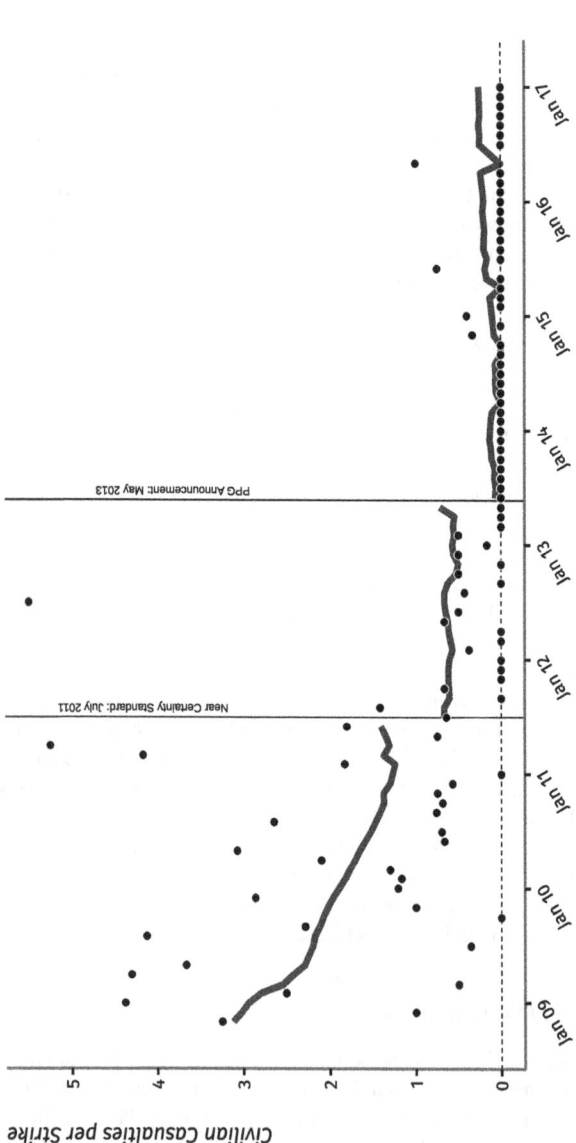

Figure 4.1 Predicted Values from Local Poisson Regressions

Note: To empirically identify a discrete change in civilian harm during U.S. drone strikes in Pakistan, we estimated local Poisson regressions that reflect the expected rate of change of civilian harm for each month between October 2010 and May 2013. The black dots represent civilian casualties from U.S. drone strikes clustered at the monthly level. The lines depict the expected rate of change for civilian casualties, also calculated at the monthly level, which declined with the adoption of the near certainty standard in July 2011 to virtually zero by May 2013, when Obama formally announced the policy change.

killed two civilians, whereas a near certainty strike killed none. Finally, Table 4.1 shows that while civilian harm was significantly lower in strikes conducted under the near certainty standard, strike frequency and combatant casualties did not see reductions of comparable magnitude. Rather, our analysis shows that the near certainty standard allowed for a similar frequency of operations while reducing civilian casualties from U.S. drone strikes by 2 per month (−72%) and increasing strike precision by 12 percentage points (18%).

Socio-Economic Implications

We draw on these findings to explore the socio-economic implications of the near certainty standard, which relies on two important assumptions. First, we assume that strikes are "as-if" random, meaning the timing of strikes is subject to chance. Rigterink (2021), for instance, shows that strike "hits" and "misses" are quasi-random. Johnston and Sarbahi (2016), as well as Abrahms and Mierau (2017), also argue that while decision makers' use of strikes may not be random, it is impossible to forecast the results of strikes. Indeed, it may actually be the case that the timing of a strike depends on the operational maintenance of a drone, that is, whether a drone has a malfunction in flight (Benjamin 2012) or variable weather conditions, including wind gusts (Mahmood & Jetter 2023). Together, these explanations align with research by Harada et al. (2022) on the randomness of bombing in war. Second, we assume that civilians do not "select into treatment," meaning they do not willingly expose themselves to strikes. Rather, the literature suggests that drones paralyze civilians with fear, which causes them to make significant adjustments to their daily patterns of life to avoid being targeted (Page & Williams 2022).

We leverage these two assumptions for civilian harm to sample and match—with replacement—casualty values for strikes approved under the reasonable certainty standard to those during the near certainty period after July 2011. We perform a Monte Carlo simulation with 5,000 iterations and average across matched civilian casualty values for each treated strike to calculate averted casualties attributable to a shift in unilateral constraint toward near certainty. We first take the difference between the matched average civilian casualty values and the BIJ-reported outcomes. We then sum up these differences to estimate the averted civilian and combatant deaths.

This calculation suggests that the policy adjustment increased the precision of U.S. strikes to a level that they only killed the intended target(s). Prior to the policy's adoption, fewer than 70% of those killed by U.S. drone strikes in Pakistan were the intended target. Following the policy's implementation, the average value of strike precision ballooned to 95%, suggesting that U.S. drone strikes in Pakistan reached "near-unerring accuracy" (Khan 2021) without imposing unintended harm on civilians. Indeed, our analysis shows that Obama's adoption of the near certainty standard in July 2011 averted 284 civilian casualties during U.S. drone strikes in Pakistan. In Figure 4.2, we plot

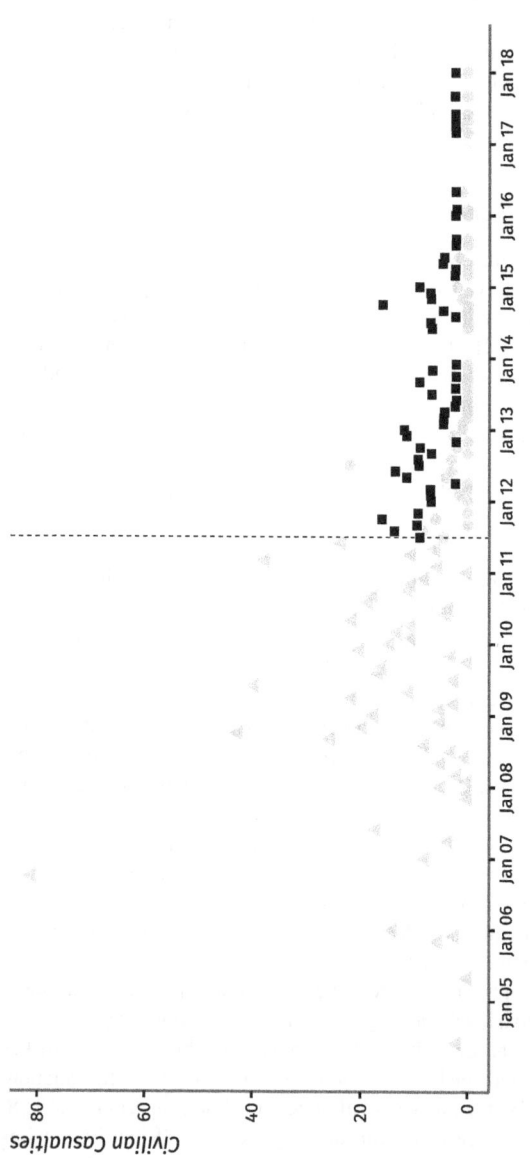

Figure 4.2 Monte Carlo Simulation of Reasonable Certainty Standard Strikes

Note: This figure shows results from our Monte Carlo simulation of expected civilian and combatant casualty outcomes had the Obama administration maintained the reasonable certainty standard. The highlighted points are the results from our simulation, with the shape of the points reflecting the certainty standard under which the strikes were conducted.

the distribution of data generated in the Monte Carlo simulation of reasonable certainty strikes and for BIJ data on strikes conducted under each certainty standard for civilian and combatant casualties. Specifically, we plotted BIJ and simulated data from 2004 to 2018 at the monthly level. This illustrates the difference in civilian casualties, accounting for changes in monthly strike frequency. The difference in outcomes does not persist for combatant casualties, as our simulated values are descriptively similar to BIJ reporting. The 284 averted civilian deaths reflect an 81% difference between simulated U.S. drone strikes observed in the BIJ dataset that were conducted with reasonable and near certainty. This outcome aligns with our strike-level estimate for the effect of near-certainty operations on civilian deaths.

These averted civilian casualties represent a large financial and psychological toll had they occurred, though scholars understand these implications much less than they do for traditional measures of civilian harm, especially injury and death. Admittedly, this toll is difficult to quantify. Some analysts may also contend that attempting to do so is morally problematic (Lushenko & Raman 2022). Despite these concerns, we attempt to offer the first socioeconomic interpretation of the effects of Obama's adoption of the near certainty standard, as these economic consequences also help to condition the perceived legitimacy of U.S. drone strikes (Lushenko 2022a). We do so by providing a value-of-statistical life (VSL) calculation based on a series of estimates. These values represent the "local trade-off rate between money and fatality risk" (Kniesner & Viscusi 2019, 1). We use widely accepted VSL values to present a range of averted monetary losses attributable to the near certainty standard (Viscusi & Masterman 2017; Rafiq 2011). We use a low-end estimate of $320,000 and a high-end estimate of $900,000 to define the VSL per individual in Pakistan. Using these figures, we calculate that the implementation of the near certainty standard averted a VSL loss of $90 to $250 million, which we interpret as a conservative estimate given the lack of transparency for U.S. drone strikes in Pakistan.

Our simulations suggest that the change in strike frequency after implementation of the near certainty standard reflects a re-calibration of U.S. drone strategy against the most important terrorists and led to 115 fewer combatant deaths. Rather than reflecting negligence, as some critics claim, we argue these averted combatant deaths did not come at an appreciable cost to U.S. national security. There were no major terrorist attacks on U.S. soil during Obama's presidency, which is the most important national security interest compelling the U.S. occupation of Afghanistan and counterterrorism operations across the region, including in Pakistan. This implies that the near certainty standard effectively balances military necessity against distinguishing between combatants and non-combatants during U.S. drone strikes in Pakistan. Still, it is possible to quantify the implications of averted combatant deaths in economic terms if we assume that each may have participated in a suicide bombing operation within Pakistan.

From 2009 to 2018, the University of Maryland's Global Terrorism Database shows that Pakistan experienced 631 suicide attacks, of which 59% were waged by vehicles and 41% by individuals (see Appendix B for a graph depicting these attacks across time) (University of Maryland 2021). Using these figures, the 115 averted combatant deaths represent the potential for an additional 68 vehicle-borne and 47 person-based suicide attacks. On the basis of values provided by the U.S. Joint Improvised Explosive Device and Defeat Organization, we assume these attacks cost $20,000 and $1,200, respectively (Ackerman 2011). This suggests that the 115 averted combatant deaths amount to a potential incurred economic cost of nearly $1.4 million that is recouped by three averted civilian deaths. At the same time, we find that the total number of suicide attacks peaked in 2013, after which it declined overtime, further suggesting that Obama's announcement of the near certainty standard may have had some effect in shaping the public's perceptions of legitimate strikes, as we discuss below.

Implications

We provide the first quasi-experimental evidence that conditioning drone strike approval on the near certainty standard prevented 284 civilian deaths in Pakistan and averted a VSL loss of more than $200 million. These findings for the near certainty standard are likely to have significant impacts on the public's perceptions of legitimate strikes, especially among Pakistani citizens and perhaps others in different targeted areas. Presently, the U.S. schema for compensation to families of civilians killed by U.S. drone strikes is not publicly known. The available evidence suggests that such U.S. compensation or "solatia" payments are wholly inadequate—even in developing countries such as Pakistan, where the earning potential of citizens is comparatively less than it is for developed states. The usual amount awarded by the U.S. government for a civilian death is $2,500 (Gilbert 2015).[1]

These payments also dramatically undervalue victims, likely contributing to grievances that stoke political violence. Silverman shows that adequate post-harm compensation reduced insurgent violence in Iraq from 2004 to 2008 because it constituted a costly signal for the U.S. government's sincere contrition. For every six-month period of conflict, Silverman (2020) finds a $2,000 increase in solatia payments prevented one more insurgent attack in Iraq, which is not an insignificant outcome given that such attacks are specifically designed to produce mass casualties (Jaeger & Siddique 2018; Shah 2018; Bergen & Tiedemann 2011). In the context of U.S. drone strikes in Somalia, Mueller (2023) also adds that U.S. officials should convert limited, infrequent, and symbolic solatia payments into a broader reparations program to offset the financial hardship and psychological toll incurred by the families of unintended victims. Doing so, Mueller (2023) contends, would help U.S. officials meet a moral obligation to more meaningfully atone for gross violations of International Humanitarian Law. Reparations may also help prevent the regeneration of terrorist groups, which are thought to recruit members on the basis of U.S. military targeting errors that result in unintended civilian casualties.

Indeed, not only does prior research suggest that Pakistani citizens largely perceive U.S. strikes as indiscriminate, which is likely to be the case in other countries similarly afflicted with high civilian death tolls following U.S. strikes, such as Iraq and Syria (Silverman 2019). Research also shows that each civilian casualty resulting from a poorly executed strike can engender 20 additional terrorists as a function of radicalization, which reflects the public's poor perception of legitimacy for drones (Gregory 2022). Taken at face value, this suggests that the 284 prevented civilian deaths in Pakistan potentially had a broader impact: averting the radicalization and recruitment of up to 6,000 Pakistani citizens and reflecting the perceived illegitimacy of over-the-horizon strikes governed by the more lenient reasonable certainty standard.

Our results highlight the importance of recalibrating targeting standards that condition military action and suggest that adopting the near certainty standard should be considered as a unilateral constraint for mitigating civilian harm during U.S. drone strikes abroad, even in declared theaters of operations that have U.S. boots deployed on the ground. More generally, our analyses highlight the disparity between the economic toll incurred by civilian harm and compensation offered to the families of victims in the form of solatia payments. We conservatively estimate that, if all victims' families were offered the usual level of compensation, U.S. solatia payments would account for just 0.7% of the $128 million VSL loss incurred by reasonable certainty drone strikes in Pakistan. Over-the-horizon strikes predicated on a unilateral constraint of near certainty may then provide the best means to shape the public's perceptions of legitimate drone warfare, both at home and abroad, which is supported by several other studies as well.

Following the policy's implementation, one Pakistani citizen reported that "[w]e never had a drone strike in our village because we didn't have any militants in our area," which Ansari interprets as a vote of "confidence in the targeting of the drone program" (Ansari 2022a, 2022b). Shah also finds, having conducted 167 interviews of informed Pakistani citizens who either live in or are otherwise related to residents of the Federally Administered Tribal Areas (FATA), that observers of U.S. drone strikes authorized under Obama's near-certainty standard thought they were legitimate. In one testimony, characteristic of similar statements, one interviewee argued, "I think most of the tribal people know that drones are precise. Those who died in drone attacks were already militants or al-Qaida and Taliban leaders" (Shah 2018, 61). In an extremely rare interview, a Pakistani general responsible for Pakistani soldiers deployed to the FATA offered similar praise, confirming the perceived legitimacy of U.S. drone strikes:

> Myths and rumors about US predator strikes and the casualty figures are many, but it's a reality that many of those being killed in these strikes are hardcore elements, a sizable number of them foreigners. Yes there are a few civilian casualties in such precision strikes, but a majority of those eliminated are terrorists, including foreign terrorist elements.
>
> (Williams 2011)

Finally, Mahmood and Jetter (2023) use wind gusts to help instrument the implications of U.S. drone strikes on Pakistani citizens' attitudes. Wind gusts constitute an exogenous variable that makes U.S. drone strikes "as-if" random, meaning the outcomes for Pakistani citizens' attitudes can be reasonably assumed to result from a natural experiment. A wind gust, in other words, is (1) not designed nor implemented by researchers, (2) is therefore unknown to researchers beforehand, and (3) there is equiprobability that U.S. drone strikes are subject to this variable weather condition (Titiunik & Sekhton 2012). Using this design-based identification strategy, known as an instrumental variable estimation, Mahmood and Jetter (2023) draw data from the leading English-language newspaper in Pakistan, *The News International*, to analyze public sentiment for U.S. drone strikes in the country. They find that positive, and to a lesser extent negative, sentiment for U.S. drone strikes increases coincident with the onset of the Obama administration's near-certainty targeting standard, reflecting a more favorable perception of the legitimacy of these operations than analysts have previously reported.

Note

1 As a result, non-governmental organizations such as Global Exchange have emerged to compensate innocent victims of U.S. drone attacks (Benjamin 2012).

5 Multilateral Constraint and Public Perceptions of Legitimacy

Our analysis in Chapter 3 suggests that French perceptions of legitimate drone warfare can also be significantly moderated by how strikes are constrained. Whereas Americans tend to privilege unilateral constraint, unless or until strikes result in civilian casualties, French citizens prefer multilateral constraint regardless of the potential for civilian harm. The purpose of this chapter is to better understand these dynamics as well as the mechanisms that may shape public preferences for multilateral constraints in terms of perceived legitimacy. We administer a survey experiment across nationally representative samples in France and the United States to determine the channels by which the public connects international approval through the UN, perhaps the most well-known multilateral constraint as we discuss in Chapter 2, with perceptions of legitimate drone warfare. We develop and test hypotheses that relate to the perceived legality, merit, and morality of drone strikes given international approval, as well as the prospects for burden-sharing, in which countries help each other offset the costs of operations abroad. This approach complements our examination in Chapter 3, wherein we vary multiple strike attributes rather than focusing on one.

Our findings in this chapter suggest that UN approval can shape the public's perceptions of legitimate drone warfare and that respondents emphasize perceived compliance with international law the most when intuiting the appropriateness of strikes. This result is strongest among U.S. respondents, which is a puzzling outcome, though it reinforces results from Chapter 3 when deliberating controlling for the type of tactical or strategic strike. Americans' preferred model of over-the-horizon strikes often transgresses other countries' sovereignty or territorial integrity. This threatens to impose what Brunstetter and Férey refer to as an "imperial slide," which others, including former U.S. President Barack Obama, have identified as a moral hazard. The record of evidence indicates that U.S. political and military officials are more willing to use drones, considering their forces do not incur as much physical risk on the battlefield. Equally ironic is Americans' emphasis on burden-sharing, especially in the case that strikes are conducted by another country while also being multilaterally constrained. Though this finding is consistent with the

DOI: 10.4324/9781032614267-5

United States' support for internationally authorized interventions conducted by its allies (Blankenship 2021), it also suggests that Americans hold their country to a different standard than others in terms of strikes, an outcome for which our results provide statistical support.

In the remainder of this final empirical chapter, we first introduce our theoretical expectations for the relationship between multilateral constraint and public perceptions of legitimate drone warfare, deliberately controlling for the type of tactical or strategic strike. We then outline our research design before presenting our results. We conclude by addressing the contributions of our findings.

The Mechanisms of Multilateral Constraint

Scholars concur that international authorization can favorably shape public opinion for countries' use of force abroad (Fang & Oestman 2022; Busby et al. 2020; Tago & Ikeda 2015). Whether the public's preferences for multilateral constraint extend to drones, and if so, how, has largely escaped academic study. This is a problem because the juridical pattern of drone warfare, which some scholars also refer to as a French model of strikes that we explained in Chapter 2, is predicated on international approval as a key form of multilateral constraint. In light of this observation, we discuss several mechanisms for international approval and how these may relate to countries' use of drone strikes.

First, international approval indicates a shared belief in the anticipated benefits of force (Voeten 2005). Strikes, therefore, are understood as less politically motivated and constitute a defensively oriented approach that promotes force as an appropriate last resort, in line with the *jus ad bellum* (just recourse to war) component of International Humanitarian Law. This indicates that the objectives of strikes have been vetted, subject to the dispassionate opinions of other member states of international society (Grieco et al. 2011). The consent and cooperation afforded by UN approval also suggest that strikes are multilaterally constrained and more compatible with social goals shared by all countries. That other, potentially more suspicious countries endorse strikes signals that they are designed to achieve common objectives, such as preventing a humanitarian crisis, interdicting terrorists, or removing a dictator. Thus, we hypothesize that UN approval can moderate the public's perceptions of legitimate strikes by signaling broad consent for the intended objectives.

Second, UN approval for countries' use of force also suggests military action is likely to comply with International Humanitarian Law. The law governing the use of force consists of *jus ad bellum* (just recourse to war) and *jus in bello* (just use of force in war) components. By violating another country's sovereignty, drones may contravene *jus ad bellum* norms of proper authority, just cause, and last resort. Poor intelligence as well as analytical biases—such as confirmation bias, wherein analysts discount contradictory evidence—may lead to botched strikes that cause civilian casualties, which suggests that drones can also flout *jus in bello* norms of distinction, proportionality, and necessity. We follow recent research by treating countries' legal commitments in terms of respect for sovereignty and the protection of civilians while using strikes (Kreps & Wallace 2016). We hypothesize that UN approval can affect the public's perceptions of legitimacy by suggesting compliance with International Humanitarian Law.

Third, previous research suggests that the public is concerned with the morality of countries' use of force abroad (Tomz & Weeks 2020). Dill and Schubiger (2021), for instance, show that the public seems to combine normative and instrumental concerns about right and wrong when adjudicating support for the use of force abroad. Drones are thought to exacerbate this tendency because they threaten morally problematic killing. For some critics, drones impose radically asymmetric violence by preserving the immunity of one party in a conflict while consolidating the liability to be harmed completely within another side (Renic 2020). Earlier research also shows that adherence to International Humanitarian Law can mediate the public's support for strikes, and the effect appears to be for normative rather than merely instrumental reasons. We posit that UN approval can shape the public's perceptions of legitimate strikes by reflecting a higher degree of morality.

Finally, scholars argue that international approval signals the prospect of greater burden-sharing (Wallace 2019, 2013). Milner and Tingley (2013) find burden-sharing diffuses the costs of providing public goods across many countries. Further, burden-sharing provides "political cover of shared blame if the operation goes awry" (Finnemore 2003). In the context of drone warfare, the most egregious mistake consists of civilian causalities. International authorization, then, could influence the public's perception of burden-sharing, which has effects on their perceived legitimacy of strikes. On the other hand, it may also be the case that burden-sharing can be less relevant for countries' use of strikes because of drones. As we discussed in the introductory chapter, for instance, political and military officials often adopt drones to minimize the financial and human costs of using military force abroad. In fact, this is the chief benefit of drones.

Data and Experimental Design

To test these expectations, we fielded an original survey experiment among approximately 900 American and 900 French citizens between November 2 and 16, 2021. Similar to the survey we used to generate results in Chapter 3, we enlisted Qualtrics to source representative panels of respondents and then blocked these against several demographics, including age, education, and gender. In this way, we resolve issues associated with other online recruitment protocols, namely Amazon Mechanical Turk. While randomized controlled trials using convenience samples may be more accessible for researchers, they are subject to the same selection bias that hounds observational studies and lack external validity (Findley et al. 2020). It is well known that respondents recruited from Amazon Mechanical Turk, for instance, are predominately younger, better educated, and more progressive than the target population. These potentially confounding variables force researchers to include controls to manage the potential for skewed results, which is not the case with representative samples.

Our survey follows a 2×2 factorial and between-subject design with four randomized prompts presented to respondents (the full survey scenario, questionnaire, and summary statistics are included in Appendix C). This means that we vary two conditions along two axes, including (1) multilateral constraint qua international approval through the UN for a strike (yes *or* no) and (2) the country approved to conduct a strike (France *or* the United States). As such, we follow existing research by designing our scenario around terrorism that constitutes a national security threat to both France and the United States (Recchia & Chu 2021). Although many political scientists present vignettes that incorporate fictional country names to manage the effects of priming and social desirability bias, as we did in Chapter 3, we opt to use scenarios that correspond to realistic examples in this case. This strategy helps us prevent bias that could shape respondents' attitudes toward a strike in terms of the location, which could be problematic for French subjects given the country's exclusive use of drones in Western Africa. Admittedly, our approach may also impose trade-offs, as we allude to in Chapter 3. Using real country names in scenarios may make respondents less likely to support the use of force abroad (Majnemer 2023; Brutger et al. 2022). In the context of this chapter's study, then, we risk distorting the moderating impact of multilateral constraint—in this case, UN approval—on the public's perceptions of legitimacy. On the other hand, the use of real country names in our vignettes protects against respondents' preexisting beliefs clouding their judgments, which has important implications for the generalizability of the results (Kreps & Roblin 2019).

Following the vignette, we probe our main dependent variable, consisting of perceived legitimacy. We ask respondents to rate their perceptions of how legitimate the strike was using a 5-point scale ranging from "very legitimate" (5) to "not legitimate" (1). We then move to questions designed to measure the public's perceptions of merit, legality, morality, and burden-sharing while also randomizing the question order to help mitigate the potential for bias (Mutz 2011). For merit, we ask respondents to rate their perceptions of the costs and benefits of the strike across six related questions and find the mean of these responses. Of note, this definition and index measurement of merit are different from our approach in Chapter 3, and the results should be interpreted accordingly. For legality, we ask respondents to judge the degree to which the strike was compatible with international law. For morality, we ask respondents about the country's moral obligation to use the strike. For burden-sharing, we ask respondents how likely it is that other countries will help carry out the strike.

Experimental Results

Overall, the results show that multilateral constraints in terms of UN approval can favorably shape the public's perceptions of a legitimate strike, which is useful to reinforce our findings from Chapter 3 while also echoing the results of previous research. Fang and Oestman (2022), for instance, find that the public perceives multilateral forms of military intervention as most legitimate, regardless of the humanitarian or security purpose.[1] In Table 5.1, we use a multivariate regression to show the causal relationship between randomized strike attributes and the outcome of the public's perceptions of legitimacy when controlling for the type of tactical or strategic strike. We observe that the public perceives strikes conducted with unilateral constraint as less legitimate than if they were multilaterally constrained, that is, if they were endorsed by the UN. Importantly, this finding is not conditioned by the strike's arbiter or the country conducting the operation, which suggests the cross-national validity of the findings.

At the country level, French and American citizens discount the perceived legitimacy of strikes that are conducted without multilateral constraint in terms of UN approval. Specifically, French respondents' perceptions of legitimacy are significantly reduced by strikes conducted by their own country that result from nothing more than internally imposed or unilateral constraints ($\beta = -0.31$, $p < 0.05$). American respondents reflect a similar reduction in the perceived legitimacy of strikes conducted by other countries with only unilateral constraints ($\beta = -0.27$, $p < 0.05$). Together, these results imply a cross-national belief in multilateralism for normative reasons, which complements our previous results as well as existing research (Kreps & Maxey 2021).

At the same time, the public tends to penalize strikes conducted by another country without UN approval, again showing the implications of multilateral constraints for the perceived legitimacy of drone warfare in a cross-national context. French respondents report lower levels of perceived legitimacy for any unilaterally constrained strike conducted by another country ($\beta = -0.26$, $p < 0.05$). The effect among French observers is similar in magnitude and significance, regardless of the arbiter. This finding is inconsistent with Americans' perceptions of legitimate drone strikes without multilateral constraint in terms of international authorization. Generally, Americans perceive their own country's strikes that are unilaterally constrained as legitimate, which tracks with findings from Chapter 3 for Americans' preferred model of drone warfare.

Together, these findings are important for several reasons. First, they reinforce our findings from Chapter 3 that suggest American and French citizens prefer unique models of drone warfare that relate to what we refer to as over-the-horizon and juridical strikes, respectively. Whereas Americans prefer strategic strikes that are unilaterally constrained, especially when the prospects for civilian harm are low, French citizens are more inclined to endorse strikes that are multilaterally constrained and used for tactical purposes, as is the

Table 5.1 Multivariate Regression Results

	Legitimacy	Merit	Burden-Sharing	Legality	Morality
Strike Attributes					
Other country, multilateral	-0.005 (0.949)	-0.071 (0.31)	0.136* (0.048)	0.213*** (0.001)	-0.001 (0.99)
Other country, unilateral	-0.253*** (0.000)	-0.1 (0.152)	0.035 (0.611)	-0.254*** (0.000)	-0.088 (0.18)
Own country, multilateral	0.014 (0.845)	-0.04 (0.567)	0.071 (0.298)	0.163* (0.011)	0.061 (0.348)
Own country, unilateral	-0.227** (0.002)	-0.077 (0.267)	-0.022 (0.753)	-0.149* (0.022)	-0.053 (0.42)
Age					
in decades	0.123*** (0.000)	0.179*** (0.000)	0.026* (0.049)	0.042*** (0.001)	0.096*** (0.000)
Education					
High school (or equivalent)	0.122 (0.11)	0.197** (0.008)	0.135 (0.063)	0.11 (0.106)	0.165* (0.017)
Some college	0.05 (0.564)	0.278*** (0.001)	0.226** (0.006)	0.108 (0.162)	0.232** (0.003)
2-year college degree	0.253* (0.013)	0.257** (0.01)	0.175 (0.072)	0.152 (0.097)	0.243** (0.009)
4-year college degree	0.242* (0.01)	0.219* (0.017)	0.084 (0.351)	0.264** (0.002)	0.127 (0.14)
Professional degree	0.198 (0.023)	0.253** (0.003)	0.251** (0.002)	0.091 (0.243)	0.171* (0.03)
Gender					
Female	-0.294*** (0.000)	-0.001 (0.991)	-0.159*** (0.001)	-0.157*** (0.000)	-0.152*** (0.001)
Other	-0.362 (0.407)	-0.368 (0.384)	-0.284 (0.493)	-0.244 (0.532)	-0.423 (0.287)

Note: P-values are indicated in parentheses. Regression estimates from the full survey sample reflect the impact on randomly assigned strike attributes compared to respondents who received no strike-specific information. This model includes controls for age, education, and gender. Also, the strike attributes, gender, and education are all factor variables; their respective baselines are control, male, and less than high school education. Statistical significance is represented by *** $p < 0.01$, ** $p < 0.05$, and * $p < 0.1$.

case in Western Africa. French officials reportedly do so to reconcile competing goals of status-seeking and human protection, which we detail in Chapter 3 as well. In this chapter, we gain additional leverage over these preferred models by focusing on the implications of multilateral constraints in terms of UN approval. Second, our results also suggest that drone strikes threaten to shape perceptions of an in-group and out-group that are strongest among U.S. respondents, again reinforcing our findings in previous chapters.

These results suggest that we need to know more about the microfoundations that shape the public's perceptions of legitimacy when respondents in a cross-national setting are confronted with variation in the multilateral constraint of drones. What, in other words, can explain how variation in the constraint attribute influences the public's perceived legitimacy of a strike? To address this outstanding question, which other scholars have emphasized as well (Fang & Oestman 2022), we investigate four potential mechanisms for legitimacy and how citizens respond to the randomization of the constraint attribute.

In Table 5.1, we use a multivariate regression to present our main results for the legitimacy outcomes in column one. We show estimates for our hypothesized mechanisms and their association with each randomized strike attribute in columns two through five. The results indicate that respondents across both countries associate strikes conducted under multilateral constraint or with international approval with higher levels of perceived compliance with international law. This effect is largely driven by American respondents, though, which aligns with our findings from Chapter 3. We previously found that French citizens do not think in terms of legality as much as other factors—including morality, the role of great powers, and preferences for the use of force abroad—when adjudicating the legitimacy of strikes, particularly their preferred model of juridical drone warfare. That Americans apparently do is a surprising finding considering U.S. citizens also perceive strategic strikes with unilateral constraint as most legitimate, even though they threaten to breach other countries' sovereignty.

This outcome also reinforces research demonstrating that while Americans may want to hold U.S. officials accountable to international law, especially when strikes kill civilians, they are more concerned with the implications of drones for their own safety (Kaag & Kreps 2014). Indeed, while Americans have endorsed U.S. drone strikes on every major national poll since 9/11, they also disapprove of drones used against U.S. citizens living in other countries, though they may be suspected of plotting terrorist attacks (Davis 2019; Davis & Newport 2013). In 2011, for instance, Obama authorized a drone strike to kill a U.S. citizen in Yemen who served as a key al-Qaeda propagandist, Anwar al-Awlaki. The strike inadvertently killed another U.S. citizen, Samir Khan, who also claimed allegiance to al-Qaeda. Their deaths, one intentional and the other incidental, resulted in public condemnation because Obama's use of a drone strike to kill them circumvented their constitutionally protected right to due process in a court of law. The operation, Objective Troy, also reinforced the presidency's reputation as "judge, jury, and executioner" in terms of drone strikes abroad (Shane 2016; Sterio 2018).

Meanwhile, French respondents perceive strikes as non-compliant with international law when they are unilaterally constrained and conducted by other countries ($\beta = -0.18$, $p < 0.05$) but not their own ($\beta = -0.12$, $p > 0.1$). This finding shows that French perceptions of legitimacy mostly relate to other countries' use of strikes in terms of the legal appeal of multilateral constraints. This outcome also helps further explain why French citizens seem to discount the legal status of strikes conducted by their own country's officials, which we show in Chapter 3, even though these strikes may best align with International Humanitarian Law. Indeed, French citizens may believe that their strikes are beyond reproach. Besides reinforcing the likelihood of a juridical—French—model of drone warfare, this finding is further suggestive of an in-group and out-group for strikes. French citizens discount the legitimacy of unilaterally constrained strikes conducted by their own government, whereas Americans do not ($\beta = -0.16$, $p > 0.1$). This finding reinforces our results in Chapter 3 that Americans are much more ethnocentric about U.S. strikes than are French citizens about their own country's strikes abroad.

Our analysis of the data is also helpful to determine the implications of burden-sharing for public perceptions of legitimacy. We find that Americans emphasize the potential for burden-sharing ($\beta = 0.28$, $p < 0.05$) when observing another country secure UN approval for strikes. We interpret this result to mean Americans believe that an ally's strikes warrant support in terms of intelligence and technical assistance, which characterizes U.S. aid to French strikes against al-Qaeda and the Islamic State in Western Africa. These results are not mirrored by French respondents. What this may also indicate, then, is that French citizens endorse multilateral constraint more to redress accusations of neocolonialism than to offset the costs of expeditionary operations. Though this finding is consistent with previous research, it also suggests the need for additional study (Recchia 2020), which Brunstetter also recognizes in terms of introducing a "drone contract" to inform France's use of strikes beyond its borders and region (Brunstetter 2022).

Implications

Since the terrorist attacks of 9/11, drones have emerged as the most common use of force among Western militaries for counterterrorism. Whereas both the United States and France have used strikes abroad, representing two of only a handful of countries that use strikes beyond their borders and regions, France conducts them with multilateral constraint reflected through UN approval. While multilateral constraint has received sustained academic scrutiny for the conventional use of force (Busby et al. 2020; Fang & Oestman 2022) and the use of drones by the United States has received considerable attention

(Lushenko et al. 2022), scholars have been comparatively silent on the cross-national application of strikes and international approval.

In light of these oversights, our findings make several contributions. First, while scholars occasionally study public support for drone strikes, they predominantly draw on American respondents to investigate the United States' use of strikes. Though other countries have embraced drones, in other words, the research has not kept pace, as we detail in the introductory chapter. Our research illuminates this trend while adding cross-national evidence on the use of force. Second, we provide insights into one of the most consequential developments for global security in the 21st century: armed and networked drones. Scholars have tackled questions of proliferation, effectiveness, and democratic accountability. Less studied is the global governance of drones and its effects on the sustainability of counterterrorism strikes in a cross-national context. Finally, we contribute to an emerging literature for the political psychology of drone use, which corresponds to a renewed interest in psychological approaches to international relations broadly (Kertzer & Tingley 2018). We bring this diverse set of literature into closer alignment by measuring the effect of multilateral constraints on the public's perceptions of legitimate drone strikes, which we also do in a cross-national context for two countries that frequently use drones for counterterrorism, France and the United States.

When deliberately controlling the type of tactical or strategic strike, our findings for the implications of multilateral constraint in terms of international approval on public perceptions of legitimacy help reinforce the results from Chapter 3. Specifically, we show that Americans tend to emphasize the legal status of strikes when interpreting the legitimacy of operations. French citizens, on the other hand, do not. Rather, they tend to discount the legality of their own operations, particularly when conducted with unilateral constraint. At the same time, French citizens seem to hold other countries to a different legal standard, at least the United States, when adjudicating the legitimacy of drone warfare. This is surprising, of course, since U.S. strikes often breach other countries' territorial integrity, though Americans also seem to endorse burden-sharing for these operations. We now turn to discussing the broader implications of our body of findings in this study for evolving patterns of drone warfare globally.

Note

1 See also Haworth (2021).

6 The Future of Public Opinion and Drone Warfare Studies

Findings

How, then, does the public form opinions about the legitimacy of drone strikes? The middle-range theory we introduce and test throughout this book provides an end-to-end framework to understand how evolving patterns of drone warfare shape individual preferences for legitimate strikes, which are not analogous to attitudes of support. We show that the public's perceptions of legitimate strikes can be a function of shifting combinations of drone use and constraints to help minimize unintended consequences. This finding is useful to explain the public's mercurial understanding of legitimacy, though strikes often have similar results, including civilian casualties, as is the case with American and French strikes in Africa. Indeed, defining drone warfare in terms of the strike attributes of use, constraint, and consequences is helpful to understand the public's perceptions of legitimacy cross-nationally.

In Chapter 3, we found that Americans and French citizens prefer distinct models of drone warfare that support our theoretical expectations. This result is reflected in our triangulation of statistical analysis of empirically derived data from an original survey experiment. Whereas Americans prefer over-the-horizon strikes, French citizens emphasize juridical strikes, which some scholars have also described as a "French model" of drone warfare. We also find that the constraint attribute can significantly moderate the public's perceptions of legitimate strikes. French citizens emphasize multilateral constraint regardless of civilian casualties. Americans emphasize multilateral constraint only in anticipation of civilian casualties; otherwise, they prefer unilateral constraint. Preferences for these models may reflect different approaches to moral reasoning that French and U.S. citizens adopt while adjudicating the appropriateness of strikes. Americans' emphasis on multilateral constraint is dependent on the likelihood of civilian casualties, which could suggest a probabilistic form of moral reasoning. French citizens, on the other hand, expect officials to demonstrate foresight in taking every precaution to protect civilians before strikes are taken, which may explain their preference for multilateral constraint regardless of the expectation of collateral damage (Kneer & Machery 2019). At

DOI: 10.4324/9781032614267-6

the same time, we identify the possibility of a "legitimacy paradox," wherein respondents' perceptions of legitimacy and attitudes of support deviate in terms of certain models of drone warfare. Through this finding, we further reflect that legitimacy matters when assessing the impact of evolving patterns of drone warfare globally.

These results encouraged us to more rigorously explore unilateral and multilateral constraints in Chapters 4 and 5. Our findings in Chapter 4 suggest that while Americans' preferred model of over-the-horizon drone warfare can impose unnecessary risks on civilians, adopting the unilateral constraint of "near" certainty of no civilian casualties can dramatically reduce the potential for civilian harm and help shape the perceived legitimacy of U.S. strikes abroad, especially among Pakistani citizens and military leaders, and perhaps among other populations in different targeted areas as well. In Pakistan, where Obama used drones prolifically, citizens celebrated drone strikes governed by the "near" certainty standard as a tool of justice, suggesting they perceived these operations as relatively legitimate. Key military commanders agreed, noting these operations were precise, enhancing their perceived legitimacy.

We explore multilateral constraint in Chapter 5, given the centrality of this attribute to French citizens' perceptions of legitimacy in terms of the juridical model of drone warfare. Our analysis largely reinforced the findings in Chapter 3. We show that French citizens tend to discount the legality of strikes that are unilaterally conducted by either their own country or the United States, and that Americans are comparatively more likely to understand the legitimacy of strikes in terms of their legality and prospects for burden-sharing. This is puzzling considering Americans' preferred model of over-the-horizon strikes, by definition, contravenes international law, namely the sovereign equality of countries. Americans, then, seem to endorse the idea of compliance with international law while preferring latitude for U.S. political and military officials to use strikes when, where, and how they like, at least until mistakes are made.

Overall, these findings help define the contours of an emerging "second generation" of research for public opinion and drone warfare. The first generation, benchmarked in part by Kreps' (2014) path-breaking work, focuses on U.S. citizens as a proxy for global public opinion, privileges a bottom-up interpretation of public opinion, and does not explicitly explore the microfoundations that underlie public attitudes. This phase of scholarship also emphasizes public attitudes in terms of approval or support, thereby eschewing perceptions of legitimacy. The second generation builds on this research to adjudicate public opinion comparatively, attempts to reconcile bottom-up and top-down explanations of public opinion by understanding drone warfare as a leader-driven practice, and abstracts microfoundations through causal mediation analysis, which shows the proportion of the treatment effect attributable to certain indirect mediators. It also promotes perceptions of legitimacy as a public attitude worthy of rigorous empirical study. While others may agree, they are reticent, for a variety of sound methodological reasons we explore in the introductory chapter, to treat legitimacy as a dependent variable. Yet researchers broadly recognize that legitimacy is foundational to sustainable policies. McDonald (2021, 539), for instance, claims legitimacy is "central" to countries' adoption of drones.

Limitations

While our results are promising to extend the second generation of research for public opinion and drone warfare, it is important to acknowledge several possible limitations of our study. In addition to several limitations we discussed in Chapter 1, we used an abstract encouragement strategy in Chapter 3 to design survey experiment vignettes as well as calibrate the contextual detail to balance experimental control with external validity (Brutger et al. 2022). Even so, social desirability bias, priming, and treatment effects are typical concerns for survey experiments. While we have no reason to believe our statistical results were distorted, especially because we adopted multiple robustness checks that validated our results, we also recognize that our innovative problem-oriented approach provides a useful platform for further research.

At the same time, methodologists may criticize our use of causal mediation analysis, though we attempted to account for their concerns by fulfilling key assumptions for the use of this method as well as casting a wide—but informed—net for likely mediators, which helps protect against over-estimating the results (MacKinnon et al. 2007). We also used a statistical approach to mediation analysis rather than a descriptive one. When used with surveys, the descriptive approach relies on researchers' judgment of respondents' answers to open-ended questions (Horowitz & Kahn 2021), as well as their mean responses to questions about potential mediators (Lin-Greenberg 2022), to deduce core beliefs and values that may underlie public attitudes. While useful, these descriptive methods expose researchers to bias, especially selection bias and confirmation bias (King et al. 1994).

Future Research

Set against our findings, we contend that scholars should address at least five questions to advance the research agenda for public opinion and drone warfare. First, what are the implications of varying civilian casualties for public perceptions of legitimate strikes? Our findings point to the need for more research to determine the moderating effects of shifting magnitudes of civilian deaths on perceptions of rightful conduct following drone strikes. To the extent existing research incorporates civilian casualties into survey vignettes, the collateral damage estimate is low, usually no more than a handful of civilian deaths. There is a good reason for this choice of treatment. According to the BIJ data on the universe of U.S. drone strikes from 2002 to 2018, the reported range of civilian casualties per strike is 0.53 to 1.15. On average, then, one civilian casualty is killed per U.S. strike. This is a tragic finding given the promised benefits of drones, which are to reduce risk to a country's own forces when surgically removing targets and protecting citizens. Yet this finding also helps justify our decision to include one civilian casualty rather than multiple non-combatant deaths in our survey experiment scenarios, which other studies often do but with little to no explanation for why (Rosendorf et al. 2022).

However, there are at least two trade-offs in keeping the civilian casualty treatment low. First, these low per-strike figures reflect considerable harm when aggregated over time, amounting to thousands of civilian deaths.[1] Second, scholars lack understanding of what is called the "risk ratio." This is defined as citizens' belief in the "acceptable" number of civilians to soldiers killed during a combat operation. While we have a good understanding that at least Americans' risk ratio for conventional—country-on-country—war lies somewhere between 4:1 and 40:1, we cannot be sure for drone warfare and in a cross-national context (Sagan & Valentino 2018; Tirman 2011). The findings are important to clarify intuitions about the moderating effect of civilian casualties on the public's perceptions of legitimate strikes. While Regan (2022, 314) posits that "avoiding civilian harm may be important to the perceived legitimacy of strikes," this is an empirical question he cautions deserves more study.[2] Regan (2022) notes, for instance, that it "could be that when force protection is not regarded as a relevant consideration, people believe that there is a greater responsibility to avoid civilian casualties."

There is the added benefit of varying civilian casualties as an independent variable in survey research. It helps reconceptualize the effectiveness of countries' use of drones. Presently, scholars—as well as policy-makers, military practitioners, and casual observers—perceive the effectiveness of strikes in terms of the number of dead terrorists, especially high-value targets, as well as the consequent reduction in terrorist attacks (Schwartz et al. 2022; Hardy & Lushenko 2012). Less attention is paid to the protection of civilians, which is a strategic, moral, and legal imperative imposed by International Humanitarian Law. While our initial research shows that civilian casualties can moderate public perceptions of legitimacy, we need to know more, which will help shape civilian protection as a measure of effectiveness for strikes among a broad audience of experts, practitioners, and citizens. The upshot of this shift in perspective, at least theoretically, is countries' heightened sensitivity to civilian casualties, perhaps leading to a more deliberate integration of unilateral and multilateral constraints to mitigate unintended civilian harms, as we suggest in Chapter 3. Indeed, Muhammedally and Mahanty (2022, 8) argue that "the protection of civilians should be integrated into [military] planning, intelligence, operations, targeting, and training, and should be considered when determining what lessons can be learned from past missions."

Fortunately, it now appears that U.S. officials recognize the legitimacy deficit of their strikes abroad and are making critical adjustments to drone policy and operations. In early October 2022, the Biden administration reintroduced the Obama administration's "near" certainty standard for no civilian casualties during strikes (Savage 2022).[3] Several months earlier, in late August 2022, the U.S. Department of Defense also released the "Civilian Harm Mitigation and Response Action Plan," or CHMR-AP, to help minimize civilian casualties during strikes (U.S. Department of Defense 2022a). The 36-page plan institutes a litany of adjustments. Among these, it defines the "civilian environment" as the context for military operations, which includes civilians and the infrastructure upon which their livelihoods depend. It introduces "Civilian Environment Teams" to help commanders understand how operations affect civilians. And it establishes "Red-Teaming"—consisting of experts that can challenge an organization's assumptions—to minimize cognitive biases, such as confirmation bias, that can result in preventable targeting errors. The CHMR-AP institutes a new architecture across the U.S. military, overseen from the Pentagon, to implement these sweeping doctrinal, planning, and training changes to mitigate civilian casualties during future wars. Aside from drones, the Secretary of Defense, Lloyd Austin (U.S. Department of Defense 2022b), emphasized that the plan is scalable to large-scale combat operations between countries as well.

These initial adjustments have also encouraged a cultural shift within the U.S. military for how officials understand civilian protection as not only a measure of performance but effectiveness as well, which is a change human rights advocates and groups have long endorsed. Instead of merely purporting that U.S. strikes are "righteous" or legitimate because of their removal of suspected terrorists, as was the case following the Biden administration's botched strike in Afghanistan in August 2021, political and military officials now explain in great detail what unilateral constraints they adopted to prevent unintended consequences. Biden's statement to the American people following the death of al-Zawahiri is one recent example of the U.S. military's heightened sensitivity to civilian casualties, given important policy guidance from political officials. He explained that the

> mission was carefully planned and rigorously minimized the risk of harm to other civilians. And one week ago, after being advised that the conditions were optimal, I gave the final approval to go get him, and the mission was a success. None of his family members were hurt, and there were no civilian casualties.
>
> (Biden 2022)[4]

Second, how do citizens within targeted countries perceive the legitimacy of strikes, which we briefly touched on in Chapter 3? This is an important question to adjudicate the merits of countervailing assumptions that underwrite drone policies globally, including perceptions of historically unprecedented precision on the battlefield and radically asymmetric violence against relatively defenseless targets. Silverman (2019) notes that understanding the attitudes of citizens in countries where strikes occur is important for two more reasons. First, civilians can shape the information environment through protests that spill over into social media and reshape the strategic context of conflict. Second, citizens' beliefs about a conflict can shape their support for one side or the other and how aid is channeled. It is puzzling, then, that scholars have not broadened the scope of their unit of analysis to include the attitudes of the public who are on the receiving end of strikes.

In fact, Silverman is one of only a handful of empirical researchers who have recognized that the existing scholarship fails to account for the attitudes of citizens within countries targeted by drone strikes. He uses a survey experiment administered among 1,000 Pakistani citizens to determine how variation in the targeting country (Pakistan *or* the United States) and consequence (civilian casualties *or* not) shape their beliefs about a strike. Silverman finds that the intervening country strongly moderates Pakistani citizens' perceptions of a fictional strike's discrimination between combatants and non-combatants. At the same time, he cautions that his core finding is conditioned by a key heterogeneous effect consisting of Pakistani citizens' level of religiosity. The more devout a Pakistani citizen is, the more he or she believes a strike is inherently indiscriminate, imposing a higher liability to be harmed among civilians.

Though these results are somewhat intuitive, Silverman's study is important for several methodological reasons. First, it is systematically executed, meaning the results are falsifiable and replicable, allowing other researchers the ability to probe the data for additional findings. Second, to the extent other scholars assess the attitudes of those targeted by strikes, they adopt ethnographic research designs that, while novel, are plagued by endogeneity, especially selection bias and reverse causation. Their respondent pools consist of citizens somehow directly or indirectly affected by drones, meaning it is difficult, if not impossible, to determine what actually accounts for citizens' beliefs (Ansari 2022; Cachelin 2022; Richardson 2022; Page & Williams 2022; Dill 2019). Is it citizens' *ex ante* beliefs about drones that shape their attitudes, however defined? Or is it citizens' exposure to harm during a strike, regardless of the intentionality, that shapes their beliefs *ex post*? Given these challenges, the task for researchers who empirically study public opinion and drone warfare is clear: use survey experiments that extend the second generation's emphasis on cross-national audiences to include citizens within countries targeted by strikes. While the results will further complicate our understanding of public opinion and drone warfare, they will be useful to justify a third question that should also help inform future research on public opinion and drone warfare.

Is public opinion for U.S. counterterrorism drone strikes racially biased? Despite strong support among Americans, some critics claim that U.S. citizens' attitudes toward U.S. counterterrorism drone strikes are racialized. Feldman (2011) contends that drones are a form of "racialization from above." This echoes Emery and Brunstetter's (2015) argument that drones constitute "aerial occupation," which other scholars interpret as a form of neo-colonialism (Cachelin 2022; Gusterson 2015; Munro 2014). For some international relations scholars, the "unambiguously" racist nature of U.S. drone strikes has also contributed to a racist hierarchy in contemporary global politics (Anievas et al. 2014). Williams claims that "drones reinforce Western-centric, racialized, and exclusionary non-Western modes of politics" (Lushenko et al. 2022, 23); Cooley (2017, 172) adds that drones "and the expansion of a supporting basing infrastructure constitute another type of hidden hierarchy in the US security network"; and Chandler (2022, 516) concludes drones "re-enact racial and colonial hierarchies through transnational networks and grounded relations that connect militarism to everyday life."

To the extent scholars have probed the relationship between race and public preferences for U.S. drone strikes, their studies are qualitative, meaning the findings are non-falsifiable, difficult to replicate, and hard to generalize. While Richardson (2020, 865) states that "racialized peoples have long been familiar with the martial gaze of drones" and Allinson (2015) characterizes drones as "technologies of racial distinctions," it is difficult to assess the validity of these claims without explicitly tapping into public attitudes. The assertion that U.S. drone strikes are racist is also liable to selection bias, as we discussed in Chapter 2. The intended targets are mostly insurgents and terrorists in war-torn countries that have darker skin. It is therefore challenging to assess the degree to which U.S. drone strikes are "racialized," promising "ever more oppressive control" over targeted communities (Richardson 2018, 92).

A myopic focus on the United States also discounts the proliferation of drones globally (Carter 2022), implying that other countries' use of strikes may be applauded by citizens that are racially biased. Officials in Turkey, for example, use drones against Kurdish separatists in Syria, resulting in accusations that the strikes are racially motivated (Rayne 2021). Thus, researchers should identify the exact mechanisms that may shape public preferences for drone strikes in terms of a heuristic or metaphorical definition of race and empirically test them, which can be done through a survey experiment. Similar to studies that examine the implications of a militarization of law enforcement (Flores-Macias & Zarkin 2022; Mummolo 2018) and humanitarian intervention (Chu & Lee 2023), researchers could use an image-based survey design to empirically assess the relationship between race and public perceptions of legitimacy when varying the skin color of a target and the geographic setting of a strike, which critics often claim shape racialized strikes (Akhter 2019).

Fourth, what are the implications of the remaining two models of drone warfare identified by our middle-range theory, aerial occupation and predatory strikes, for the public's perceptions of legitimacy? As an initial probe of legitimacy outcomes in the context of drones, we deliberately calibrated our analysis against patterns of strikes adopted by two Western great powers, France and the United States, that habitually use drones beyond their immediate borders and regions. In large part, this unit of analysis also reflects an important methodological choice. As discussed in Chapter 1, the data for at least U.S. drone strikes is better gathered and curated than it is for other countries' use of drones, notwithstanding reporting inconsistencies across datasets that we also explain in Chapter 4. American and French citizens are also more easily polled than are people in different countries and regions across the globe, especially those affected by conflict that is less conducive to the experimental approach we exercise. Still, an outstanding question remains for how emerging models of drone warfare other than those adopted by France and the United States may moderate the public's perceptions of legitimate strikes. This is an important matter of inquiry because hundreds of countries and non-state actors are adopting and using drones in ways that deviate from the patterns we study here, yet we do not know how the public perceives their legitimacy. An emerging line of experimental research studies public attitudes toward targeting in the context of renewed interstate conflict, which we recommend scholars bridge with drone warfare (Dill et al. 2023; Knuppe et al. 2023; Lupu & Wallace 2022).

Finally, scaling to the international level, what are the implications of evolving patterns of drone warfare on the public's perceptions of the legitimacy of the global order? Most research conceives of public attitudes toward drone strikes in terms of national security. This reflects earlier drone warfare scholarship that fails to engage the broader implications of strikes for international security, peace, and prosperity. A new wave of research, which some scholars have referred to as a "fourth wave," is designed to adjudicate the implications of countries' use of drone strikes for the perceived legitimacy of global order. The preceding three waves explored the drivers of drone proliferation, the effectiveness of drones, and the legal, moral, and ethical implications of strikes (Lushenko et al. 2022).

Though a contestable concept itself, we can understand global order as the pattern of intersubjectively shared norms and issue-specific institutions shared among countries that ameliorate the potential for interstate conflict and encourage cooperation for common goals, including security, stability, and prosperity. Scholars posit that countries' adoption of drone warfare can affect the normative and material pillars of global order. These include (1) hierarchical international society that structures global politics, (2) sovereignty that is the anticipated dividend of global order, (3) international law that is the key institutional form of global order, and (4) the diffusion of military capabilities that contributes to countries' adoption of coercive foreign policies. Specifically, drones can impose contradictions and complementaries within and across these four pillars that threaten to delegitimize global order. While this contribution is theoretically generative, fostering a new branch of research, it is also interpretative and non-falsifiable. Though this is by design in the spirit of theory-building, it suggests the need to use survey experiments to empirically test the degree to which drone strikes, as intuited by citizens in both targeting and targeted countries, actually implicate the legitimacy of global order.

Researchers can adopt at least three approaches to test the intuition that drones can both reinforce and erode the legitimacy of global order. They can ask respondents in cross-national settings to gauge their perceptions of the legitimacy of global order after reading a randomized vignette that varies how a strike affects a pillar—or pillars—of global order. Researchers can also conceptualize the pillars of global order as microfoundations to determine the degree to which respondents attribute shifts in their perceptions of the legitimacy of global order to these underlying considerations after reading a randomized vignette for a drone strike.

Alternatively, researchers can use a conjoint survey experiment to investigate the independent effects of preferences on numerous features of a complicated topic like a drone strike (Bansak et al. 2023). A conjoint survey outputs two key statistics: marginal means and the average marginal component effect. Marginal means constitute specific values for an attribute under study. The average marginal component effect captures how the value of a feature, in this case the pillars of global order, varies according to respondents' interpretations of a randomized strike vignette. Stated differently, it is the probability associated with respondents' preference for a certain outcome averaged over all respondents (Clayton et al. 2023; Leeper et al. 2020). Either way, these five questions on civilian casualties, attitudes of targeted communities, racial bias, remaining models of drone warfare, and global order suggest that the research agenda for public opinion and drone warfare is likely to evolve, especially as scholars come to understand the value of empirically testing public attitudes for strikes in terms of perceptions of legitimacy, the case for which we attempted to make in this book.

Notes

1 We want to thank Amelia Arsenault for this helpful observation.
2 Ford and Hoskins agree but are definitive in their assessment. They argue "the (in) visibility of civilian deaths . . . is central to the battle over legitimacy" (2022, 31).
3 Some argue that this adjustment appears to retain a loophole for the use of drone strikes by tactical commanders without presidential approval in terms of "collective self-defense," which is not technically a legal term (see Hathaway 2022).
4 Some, however, argue the CHMR-AP is less than meets the eye (e.g., Moyn 2022; Lushenko 2022b; Kreps 2022).

Bibliography

Abrahms, Max & Jochen Mierau. 2017. "Leadership Matters: The Effects of Targeted Killings on Militant Group Tactics." *Terrorism and Political Violence* 29 (5): 830–51.

Ackerman, Spencer. 2011. "$265 Bomb, $300 Billion War: The Economics of the 9/11 Era's Signature Weapon." *Wired*, September 8, 2011. www.wired.com/2011/09/ied-cost/.

Aikins, Matthieu, Christoph Koettl, Evan Hill, Eric Schmitt, Ainara Tiefenthaler, & Drew Jordan. 2021. "In U.S. Drone Strike, Evidence Suggests No ISIS Bomb." *The New York Times*, September 10, 2021. www.nytimes.com/2021/09/10/world/asia/us-air-strike-drone-kabul-afghanistan-isis.html.

Akhter, Majed. 2019. "The Proliferation of Peripheries: Militarized Drones and the Reconfiguration of Global Space." *Progress in Human Geography* 43 (1): 64–80.

Alexander, David. 2013. "Retired General Cautions Against Overuse of 'Hated' Drones." *Reuters*, January 7, 2013. www.reuters.com/article/us-usa-afghanistan-mcchrystal-idUSBRE90608O20130107.

Allinson, Jamie. 2015. "The Necropolitics of Drones." *International Political Sociology* 9 (2): 113–27.

Anievas, Alexander, Nivi Manchanda, & Robbie Shilliam, eds. 2014. *Race and Racism in International Relations*. London: Routledge.

Ansari, Neha. 2022a. "From 'God's Wrath' to 'Miracle Birds': The Increasing Acceptance of U.S. Drone Strikes in FATA." PhD Diss., Tufts University. www.proquest.com/openview/21c3e8c1574ed4624b475eca5249904b/1?pq-origsite=gscholar&cbl=18750&diss=y.

Ansari, Neha. 2022b. "Precise and Popular: Why People in Northwest Pakistan Support Drones." *War on the Rocks*, August 19, 2022. https://warontherocks.com/2022/08/precise-and-popular-why-people-in-northwest-pakistan-support-drones/.

Aslam, Wali. 2013. *The United States and Great Power Responsibility in International Society: Drones, Rendition, and Invasion*. New York: Routledge.

Baily, Michael A. 2017. *Real Econometrics: The Right Tools to Answer Important Questions*. Oxford: Oxford University Press.

Bain, William. 2023. *Political Theology of International Order*. Oxford: Oxford University Press.

Baldor, Lolita C. 2011. "Panetta Spills a Little on Secret CIA Drones." *Associated Press*, October 7, 2011. http://archive.boston.com/news/nation/washington/articles/2011/10/07/panetta_spills____a_little____on_secret_cia_drones/.

Banka, Andris & Adam Quinn. 2018. "Killing Norms Softly: US Targeted Killing, Quasi-secrecy and the Assassination Ban." *Security Studies* 27 (4): 665–703.

Bansak, Kirk, Jens Hainmueller, Daniel J. Hopkins, and Teppei Yamamoto. 2023. "Using Conjoint Experiments to Analyze Election Outcomes: The Essential Role of the Average Marginal Component Effect." *Political Analysis* 31(4): 500–18.

Barela, Steven J., ed. 2015. *Legitimacy and Drones: Investigating Legality, Morality, and Efficacy of UCAVs*. London: Routledge.

Baron, Reuben M. and David A. Kenny. 1986. "The Moderator-Mediator Variable Distinction in Social Psychological Research." *Journal of Personality and Social Psychology* 51, no. 6: 1173–82.

Bartels, Daniel, Christopher W. Bauman, Fiery A. Cushman, David A. Pizarro, & A. Peter McGraw. 2015. "Moral Judgment and Decision Making." In *Blackwell Reader of Judgment and Decision Making*, edited by Gideon Keren & George Wu, 478–515. Oxford: Blackwell.

Beetham, David. 1991. *The Legitimation of Power*. London: Humanities Press International, Inc.

Benjamin, Medea. 2012. *Drone Warfare: Killing by Remote Control*. New York: OR Books.

Bergen, Peter & Katherine Tiedemann. 2011. "Washington's Phantom War: The Effects of the U.S. Drone Program in Pakistan." *Foreign Affairs* 90 (4): 12–18.

Berinsky, Adam J., Gregory A. Huber, & Gabriel Lenz. 2012. "Evaluating Online Labor Markets for Experimental Research: Amazon.com's Mechanical Turk." *Political Analysis* 20 (3): 351–68.

Biden, Joseph. 2022. "Remarks by President Biden on a Successful Counterterrorism Operation in Afghanistan." *The White House*, August 1, 2022. www.whitehouse.gov/briefing-room/speeches-remarks/2022/08/01/remarks-by-president-biden-on-a-successful-counterterrorism-operation-in-afghanistan/.

Biegon, Rubrick, Vladimir Rauta, & Tom F.A. Watts. 2021. "Remote Warfare—Buzzword of Buzzkill." *Defence Studies* 21 (4): 427–46.

Biegon, Rubrick & Tom F.A. Watts. 2022. "Remote Warfare and the Retooling of American Primacy." *Geopolitics* 27 (3): 948–71.

Binder, Martin & Monika Heupel. 2021. "The Politics of Legitimation in International Organizations." *Journal of Global Security Studies* 6 (3): 1–18.

Biswas, Shampa. 2014. *Nuclear Desire: Power and Postcolonial Nuclear Order*. Minneapolis, MN: University of Minnesota Press.

Blakeley, Ruth. 2021. "Drones, State Terrorism and International Law." *Critical Studies on Terrorism* 11 (2): 321–41.

Blank, Lauren R. 2023. "Analyzing the Legality and Effectiveness of U.S. Targeted Killing." *Journal of National Security Law and Policy* 13 (2): 259–82.

Blankenship, Brian. 2021. "The Price of Protection: Explaining Success and Failure of US Alliance Burden-Sharing Pressure." *Security Studies* 30 (5): 691–724.

Boddery, Scott S. & Graig R. Klein. 2021. "Presidential Use of Diversionary Drone Force and Public Support." *Research and Politics* 8 (2): 1–7.

Bowen, Tyler, Michael A. Goldfien, & Matthew H. Graham. 2023. "Public Opinion an Nuclear Use: Evidence from Factorial Experiments." *The Journal of Politics* 85 (1): 345–50.

Boyle, Michael J. 2020. *The Drone Age: How Drone Technology Will Change War and Peace*. Oxford: Oxford University Press.

Braun, Christian N. 2023. *Limited Force and the Fight for the Just War Tradition*. Washington, DC: Georgetown University Press.

Brooks, Stephen G. & William C. Wohlforth. 2008. *World Out of Balance: International Relations and the Challenge of American Primacy*. Princeton, NJ: Princeton University Press.

Brunstetter, Daniel R. 2021. *Just and Unjust Uses of Limited Force: A Moral Argument with Contemporary Illustrations.* Oxford: University of Oxford Press.

Brunstetter, Daniel R. 2022. "French Drone Use in the Sahel and the Requirement of Consent." *Working Paper.* Irvine, CA: University of California, Irvine.

Brunstetter, Daniel R. & Amélie Férey. 2022. "Armed Drones and Sovereignty: The Arc of Strategic Sovereign Possibilities." In *Drones and Global Order: Implications of Remote Warfare for International Society,* edited by Paul Lushenko, Srinjoy Bose, & William Maley, 137–55. New York: Routledge.

Brutger, Ryan, Joshua D. Kertzer, Jonathan Renshon, Dustin Tingley, & Chagai M. Weiss. 2022. "Abstraction and Detail in Experimental Design." *American Journal of Political Science*: 1–16.

Buchanan, Allen & Robert O. Keohane. 2006. "The Legitimacy of Global Governance Institutions." *Ethics & International Affairs* 20 (4): 405–37.

Bukovansky, Mlada. 2002. *Legitimacy and Power Politics: The American and French Revolutions in International Political Culture.* Princeton, NJ: Princeton University Press.

Bull, Hedley. 1966. "International Theory: The Case for a Classical Approach." *World Politics* 18 (3): 361–77.

Bull, Hedley. 1975. "New Directions in International Relations Theory." *International Studies* 14 (2): 279.

Bull, Hedley. 1977. *The Anarchical Society: A Study of Order in World Politics.* London: Palgrave.

Busby, Joshua, Craig Kafura, Jonathan Monten, & Jordan Tama. 2020. "Multilateralism and the Use of Force: Experimental Evidence on the Views of Foreign Policy Elites." *Foreign Policy Analysis* 16 (1): 118–29.

Caato, Bashir M. 2022. "Somalia: Turkey's Bayraktar TB2 Drones Join Offensive Against Al-Shabaab." *Middle East Eye,* October 1, 2022. www.middleeasteye.net/news/somalia-turkey-bayraktar-tb2-drones-join-offensive-shabab.

Cachelin, Shala. 2022. "The U.S. Drone Programme, Imperial Air Power and Pakistan's Federally Administered Tribal Areas." *Critical Studies on Terrorism* 15 (2): 441–62.

Calcara, Antonio, Andrea Gilli, Mauro Gilli, Raffaele Marchetti, & Ivan Zaccagnini. 2022c. "Why Drones Have Not Revolutionized War: The Enduring Hider-Finder Competition in Air Warfare." *International Security* 46 (4): 130–71.

Calcara, Antonio, Andrea Gilli, Mauro Gilli, & Ivan Zaccagnini. 2022a. "Air Defense and the Limits of Drone Technology." *Lawfare,* July 3, 2022. www.lawfareblog.com/air-defense-and-limits-drone-technology.

Calcara, Antonio, Andrea Gilli, Mauro Gilli, & Ivan Zaccagnini. 2022b. "Will the Drone Always Get Through? Offensive Myths and Defensive Realities." *Security Studies* 31 (5): 791–825.

Calin, Costel & Brandon Prins. 2015. "The Sources of Presidential Foreign Policy Decision Making: Executive Experience and Militarized Interstate Conflict." *International Journal of Peace Studies* 20 (1): 17–34.

Callamard, Agnes. 2020. *Targeted Killings Through Armed Drones and the Case of Iranian General Qassem Soleimani.* New York: United Nations.

Callamard, Agnes & James Rogers. 2020. "We Need a New International Accord to Control Drone Proliferation." *Bulletin of Atomic Scientists,* December 1, 2020. https://thebulletin.org/2020/12/weneed-a-new-international-accord-to-control-drone-proliferation.

Canes-Wrone, Brandice, William G. Howell, & David E. Lewis. 2008. "Toward a Broader Understanding of Presidential Power: A Reevaluation of the Two Presidencies Thesis." *Journal of Politics* 70 (1): 1–16.

Cannon, Brendon. 2020. "Armed Drone Strikes and the Security of Somalia's Federal Government." *Small Wars & Insurgencies* 31 (4): 773–800.

Carr, E.H. 1949. *The Twenty Years' Crisis 1919–1939: An Introduction to the Study of International Relations*. London: Macmillan.

Carter, Keith L. 2022. "Coming Soon to a Theater (of War) Near You: Drones of All Shapes and Sizes." In *Drones and Global Order: Implications of Remote Warfare for International Society*, edited by Paul Lushenko, Srinjoy Bose, & William Maley, 191–209. New York: Routledge.

Carvin, Stephanie. 2015. "Getting Drones Wrong." *The International Journal of Human Rights* 19 (2): 127–41.

Ceccoli, Stephen & John Bing. 2018. "Taking the Lead? Transatlantic Attitudes Toward Lethal Drone Strikes." *Journal of Transatlantic Studies* 16 (3): 247–71.

Challans, Timothy L. 2007. *Awakening Warrior: Revolution in the Ethics of Warfare*. Albany, NY: State University of New York Press.

Chamayou, Gregoire. 2013. *A Theory of the Drone*. New York: The New Press.

Chandler, Katherine. 2022. "Apartheid Drone: Infrastructure of Militarism and the Hidden Genealogies of the South African Seek." *Social Studies of Science* 52 (4): 512–35.

Chapa, Joseph O. 2022. *Is Remote Warfare Moral? Weighing Issues of Life + Death from 7,000 Miles*. New York: PublicAffairs.

Chaudoin, Stephen, Brian J. Gaines, & Avital Livny. 2021. "Survey Design, Order Effects, and Causal Mediation Analysis." *The Journal of Politics* 83 (4): 1851–56.

Chavez, Kerry. 2023. "Learning on the Fly: Drones in the Russian-Ukrainian War." *Arms Control Today* 53: 1–8.

Chavez, Kerry & Ori Swed. 2021. "The Proliferation of Drones to Violent Nonstate Actors." *Defence Studies* 21 (1): 1–24.

Chavez, Kerry & Ori Swed. 2023. "The Empirical Determinants of Violent Nonstate Actor Drone Adoption." *Armed Forces & Society*. https://doi.org/10.1177/00953 27X231164570.

Chong, Dennis & James N. Druckman. 2007. "Framing Theory." *Annual Review of Political Science* 10: 103–26.

Chu, Jonathan and Carrie Lee. 2023. "Race, Religion, and American Support for Humanitarian Intervention." *Journal of Conflict Resolution*. https://journals.sagepub.com/doi/10.1177/00220027231214716.

Clark, Ian. 2005. *Legitimacy in International Society*. Oxford: Oxford University Press.

Clausewitz, Carl von. 1984. *On War*. Edited & Translated by M. Howard & P. Paret. Princeton, NJ: Princeton University Press.

Clayton, Katherine, Yusaku Horiuchi, Aaron R. Kaufman, Gary King, & Mayya Komisarchik. 2023. "Correcting Measurement Error Bias in Conjoint Survey Experiments." *Working Paper*. https://gking.harvard.edu/conjointE.

Cook, Martin. 2015. "Drone Warfare and Military Ethics." In *Drones and the Future of Armed Conflict: Ethical, Legal, and Strategic Implications*, edited by David Cortright, Rachel Fairhurst, & Kristen Wall, 46–63. Chicago, IL: The University of Chicago Press.

Cooley, Alex. 2017. "Command and Control? Hierarchy and the International Politics of Foreign Military Bases." In *Hierarchies in World Politics*, edited by Ayşe Zarakol, 154–74. Cambridge: Cambridge University Press.

Cortright, David, Rachel Fairhurst, & Kristen Wall, eds. 2015. *Drones and the Future of Armed Conflict: Ethical, Legal, and Strategic Implications*. Chicago, IL: University of Chicago Press.

Crawford, Neta. 2003. *Accountability for Killing: Moral Responsibility for Collateral Damage in America's Post-9/11 Wars*. Oxford: Oxford University Press.

Cronin, Audrey K. 2020. *Power to the People: How Open Technological Innovation Is Arming Tomorrow's Terrorists*. Oxford: Oxford University Press.

Crosby, Alfred W. 2002. *Throwing Fire: Projectile Technology Through History*. Cambridge: Cambridge University Press.

Dafoe, Allan, Baobao Zhang, & Devin Caughey. 2018. "Information Equivalence in Survey Experiments." *Political Analysis* 26: 399–416.

Davis, Alyssa & Frank Newport. 2013. "In U.S., 65% Support Drone Attacks on Terrorists Abroad." *Gallup*, March 25, 2013. https://news.gallup.com/poll/161474/support-drone-attacks-terrorists-abroad.aspx.

Davis, C.T. 2019. "Morality as Causality: Explaining Public Opinion on US Government Drone Strikes." PhD Diss., Arizona State University. https://keep.lib.asu.edu/_flysystem/fedora/c7/211447/Davis_asu_0010E_18615.pdf.

Dellmuth, Lisa & Jonas Tallberg. 2023. *Legitimacy Politics: Elite Communication and Public Opinion in Global Governance*. Cambridge: Cambridge University Press.

Demmers, Jolle & Lauren Gould. 2020. "The Remote Warfare Paradox: Democracies, Risk Aversion and Military Engagement." *E-International Relations*, June 20, 2020. www.e-ir.info/2020/06/20/the-remote-warfare-paradox-democracies-risk-aversion-and-military-engagement/.

DeVore, Marc R. 2020. "Reluctant Innovators? Inter-organizational Conflict and the U.S.A.'s Route to Becoming a Drone Power." *Small Wars & Insurgencies* 31 (4): 701–29.

Dewan, Khalil. 2021. "New Report Examines Civilian Deaths in French Drone-Instigated Air Strike in Mali." *Drone Wars*, July 22, 2021. https://dronewars.net/2021/07/22/new-report-examines-civilian-deaths-in-french-drone-instigated-air-strike-in-mali/.

Dill, Janina. 2015. *Legitimate Targets? Social Construction, International Law and US Bombing*. Cambridge: Cambridge University Press.

Dill, Janina. 2019. "Distinction, Necessity, and Proportionality: Afghan Civilians' Attitudes Toward Wartime Harm." *Ethics & International Affairs* 33 (3): 315–42.

Dill, Janina, Scott D. Sagan, & Benjamin A. Valentino. 2022. "Kettles of Hawks: Public Opinion on the Nuclear Taboo and Noncombatant Immunity in the United States, United Kingdom, France, and Israel." *Security Studies* 31 (1): 1–31.

Dill, Janina & Livia I. Schubiger. 2021. "Attitudes Toward the Use of Force: Instrumental Imperatives, Moral Principles, and International Law." *American Journal of Political Science* 65 (3): 612–33.

Dill, Jania, Marine Howlett, and Carl Müller-Crepon. 2023. "At Any Cost: How Ukrainians Think about Self-Defense Against Russia." *American Journal of Political Science*. https://onlinelibrary.wiley.com/doi/10.1111/ajps.12832

Dorsey, Jessica & Nilza Amaral. 2021. *Military Drones in Europe: Ensuring Transparency and Accountability*. London: Chatham House.

Drezner, Daniel. 2021. "Power an International Relations: A Temporal View." *European Journal of International Relations* 27 (1): 28–52.

Drone Industry Insights. 2021. *South African Drone Market Report 2019–2024: The South African Commercial Drone Market Size and Forecast 2019–2024*. Hamburg: Drone Industry Insights.

Dudziak, Mary L. 2012. *War Time: An Idea, Its History, Its Consequences*. Oxford: Oxford University Press.

Dunne, Tim. 1998. *Inventing International Society: A History of the English School*. London: Macmillan Press Ltd.

Elish, M.C. 2017. "Remote Spilt: A History of US Drone Operations and the Distributed Labor of War." *Science, Technology, & Human Values* 42 (6): 1100–31.

Emery, John R. & Daniel R. Brunstetter. 2015. "Drones as Aerial Occupation." *Peace Review* 27 (4): 424–31.

Enemark, Christian. 2023. *Moralities of Drone Warfare*. Edinburgh: Edinburgh University Press.

Engelbrecht, Cora & Euan Ward. 2022. "The Killing of Ayman al-Zawahiri: What We Know." *The New York Times*, August 2, 2022. www.nytimes.com/2022/08/02/world/asia/al-qaeda-al-zawahri-killing.html.

Erickson, Jennifer I.. 2018. *Dangerous Trade: Arms Exports, Human Rights, and International Reputation*. New York: Columbia University Press.

Fair, Christine & Ali Hamza. 2016. "From Elite Consumption to Popular Opinion: Framing of the US Drone Program in Pakistan Newspapers." *Small Wars & Insurgencies* 27 (4): 578–607.

Fang, Songying & Jared Oestman. 2022. "The Limit of American Public Support for Military Intervention." *Armed Forces & Society*: 1–24. https://doi.org/10.1177/0095327X221107700.

Fazal, Tanisha M. 2018. *Wars of Law: Unintended Consequences in the Regulation of Armed Conflict*. Ithaca, NY: Cornell University Press.

Fearon, James D. 1995. "Rationalist Explanations for War." *International Organization* 49 (3): 379–414.

Feldman, Keith P. 2011. "Empire's Verticality: The Af/Pak Frontier, Visual Culture, and Racialization from Above." *Comparative American Studies an International Journal* 9 (4): 325–41.

Ferl, Anna-Katharina. 2023. "Imaging Meaningful Human Control: Autonomous Weapons and the (De-) Legitimation of Future Warfare." *Global Society*: 1–17. https://doi.org/10.1080/13600826.2023.2233004.

Findley, Michael G., Kyosuke Kikuta, & Michael Denly. 2020. "External Validity." *Annual Review of Political Science* 24: 365–93.

Finnemore, Martha. 2003. *The Purpose of Intervention: Changing Beliefs about the Use of Force*. Ithaca, NY: Cornell University Press.

Fisk, Kerstin, Jennifer L. Merolla, & Jennifer M. Ramos. 2019. "Emotions, Terrorist Threat, and Drones: Anger Drives Support for Drone Strikes." *Journal of Conflict Resolution* 63 (4): 976–1000.

Flores-Macias, Gustavo & Jessica Zarkin. 2022. "Militarization and Perceptions of Law Enforcement in the Developing World: Evidence from a Conjoint Experiment in Mexico." *British Journal of Political Science* 52 (3): 1377–97.

Ford, Matthew & Andrew Hoskins. 2022. *Radical War: Data, Attention, and Control in the 21st Century*. Oxford: Oxford University Press.

France 24. 2021. "In Mali, France's Defense Chief Defends January Airstrike that UN Report Says Killed 19 Civilians." *France 24*, February 4, 2021. www.france24.com/en/africa/20210402-in-mali-france-s-defence-chief-defends-january-airstrike-that-un-report-says-killed-19-civilians.

France 24. 2022. "French Army Says Senior Al Qaeda Leader Killed in Mali." *France 24*, March 3, 2022. www.france24.com/en/africa/20220307-french-army-says-senior-al-qaeda-leader-killed-in-mali.

Galliott, Jai. 2015. *Military Robots: Mapping the Moral Landscape*. New York: Routledge.

Garbino, Henrique. 2023. "Rebels against Mines? Legitimacy and Restraint on Landmine Use in the Philippines." *Security Studies*: 1–32.

Gelpi, Christopher. 2003. *The Power of Legitimacy: Assessing the Role of Norms in Crisis Bargaining*. Princeton, NJ: Princeton University Press.

George, Stephen. 1976. "The Reconciliation of the 'Classical' and 'Scientific' Approaches to International Relations." *Millennium* 5: 28–40.

Gilbert, Emily. 2015. "The Gift of War: Cash, Counterinsurgency, and 'Collateral Damage'." *Security Dialogue* 46 (5): 403–21.

Gilli, Andrew & Mauro Gilli. 2016. "The Diffusion of Drone Warfare? Industrial, Organizational, and Infrastructural Constraints." *Security Studies* 25 (1): 50–84.

Goddard, Stacie E. & Ronald R. Krebs. 2015. "Rhetoric, Legitimation, and Grand Strategy." *Strategic Studies* 24(1): 5–36.

Grafstein, Robert. 1981. "The Failure of Weber's Conception of Legitimacy: Its Causes and Implications." *The Journal of Politics* 43 (2): 456–72.

Grant, Ruth W. & Robert O. Keohane. 2005. "Accountability and Abuses of Power in World Politics." *American Political Science Review* 99 (1): 29–43.

Gregory, Thomas. 2022. "Calibrating Violence: Body Counts as a Weapon of War." *European Journal of International Security* 7 (4): 479–507.

Grieco, Joseph M., Christopher Gelpi, Jason Reifler, & Peter D. Feaver. 2011. "Let's Get a Second Opinion: International Institutions and American Public Support for War." *International Studies Quarterly* 55 (2): 563–83.

Gusterson, Hugh. 2014. "Toward an Anthropology of Drones: Remaking Space, Time, and Valor in Combat." In *The American Way of Bombing: Changing Ethical and Legal Norms, from Flying Fortresses to Drones*, edited by Matthew Evangelista & Henry Shue, 191–206. Ithaca, NY: Cornell University Press.

Gusterson, Hugh. 2015. *Drone: Remote Control Warfare*. Cambridge, MA: The MIT Press.

Gusterson, Hugh. 2019. "Drone Warfare in Waziristan and the New Military Humanism." *Current Anthropology* 60 (S19): S77–86.

Hamming, Tore R. 2023. *Over-the-Horizon Counterterrorism: Implications of the New Western Approach*. New York: Dorf Publishing.

Hansen, Stig J. 2023. "Can Somalia's New Offensive Defeat al-Shabaab?" *CTC Sentinel* 16 (1): 19–24.

Harada, Masataka, Gaku Ito, & Daniel M. Smith. 2022. "Destruction from Above: Long-Term Legacies of the Tokyo Air Raids." *SSRN*, August 8, 2022. https://ssrn.com/abstract=3471361.

Hardy, John & Paul Lushenko. 2012. "The High Value of Targeting: A Conceptual Model for Using HVT against a Networked Enemy." *Defence Studies* 12 (3): 413–33.

Hartig, Luke. November 5, 2021. Personal communication with the authors.

Hartig, Luke. 2022. "The Biden Drone Playbook: The Elusive Promise of Restrained Counterterrorism." *Just Security*, October 17, 2022. www.justsecurity.org/83586/assessing-bidens-counterterrorism-rules/.

Hathaway, Oona A. 2022. "Biden's New Counterterrorism Policy Guidance Further Entrenches the Forever War." *Just Security*, October 11, 2022. www.justsecurity.org/83487/bidens-new-counterterrorism-policy-guidance-further-entrenches-forever-war/.

Hathaway, Oona A. 2023. "How the Erosion of U.S. War Powers Constraints Has Undermined International Law Constraints on the Use of Force." *Harvard Law School National Security Journal* 14 (2): 336–66.

Haun, Phil M. 2022. "Air Power in the Age of Primacy: Air Warfare since the Cold War." In *Air Power in the Age of Primacy: Air Warfare Since the Cold War*, edited by Phil M. Haun, Colin F. Jackson, & Timothy P. Schultz, 1–25. Cambridge: Cambridge University Press.

Haworth, Alida. 2021. "It's Up for Debate: A Study of Political Rhetoric, Consensus, and the Legitimacy of the Use of Force." Master's Thesis, University of Oxford.

Hazelton, Jacquelin. 2017. "Drone Strikes and Grand Strategy: Toward a Political Understanding of the Uses of Unmanned Aerial Vehicle Attacks in US Security Policy." *Journal of Strategic Studies* 40 (12): 68–91.

Herrmann, Richard, Philip E. Tetlock, & Matthew N. Diascro. 2001. "How Americans Think About Trade: Reconciling Conflict among Money, Power, and Principles." *International Studies Quarterly* 45 (2): 191–218.

Hobbes, Thomas. 1994. *Leviathan: With Selected Variants from the Latin Edition of 1668*. Edited by Edwin Curley. Indianapolis, IN: Hackett Publishing.

Hodges, Doyle K. 2018. "Let Slip the Laws of War! Legalism, Legitimacy, and Civil-Military Relations." PhD Diss., Princeton University. https://dataspace.princeton.edu/handle/88435/dsp01ns064879r.

Holeindre, Jean-Vincent. 2018. "A Certain Idea of Grandeur: French Military Interventionism and Postcolonial Responsibility." In *The Ethics of War and Peace Revisited*, edited by Daniel R. Brunstetter & Jean-Vincent Holeindre, 139–58. Washington, DC: Georgetown University Press.

Hollande François. 2013. "Hollande: 750 Hommes Au Mali, Nouvelles Frappes Réussies." *L'OBS*, January 15, 2013. http://tempsreel.nouvelobs.com/politique/20130115.AFP0522/hollande-750-hommes-au-mali-nouvelles-frappes-reussies.html.

Hopkins, Nick. 2013. "Former NSA Chief: Western Intelligence Agencies Must Be More Transparent." *The Guardian*, September 30, 2013. www.theguardian.com/world/2013/sep/30/nsa-director-intelligence-public-support.

Horowitz, Michael C. 2016. "Public Opinion and the Politics of the Killer Robots Debate." *Research and Politics* 3 (1): 1–8.

Horowitz, Michael C. & Lauren Kahn. 2021. "What Influences Attitudes about Artificial Intelligence Adoption: Evidence from U.S. Local Attitudes." *PLoS ONE* 16 (10): 1–20.

Horowitz, Michael C., Lauren Kahn, Julia Macdonald, & Jacquelyn Schneider. 2023. "Adopting AI: How Familiarity Breeds Both Trust and Contempt." *AI & Society*: 1–16.

Horowitz, Michael C., Sarah E. Kreps, & Matthew Fuhrmann. 2016. "Separating Fact from Fiction in the Debate Over Drone Proliferation." *International Security* 41 (2): 7–42.

Horowitz, Michael C. & Erik Lin-Greenberg. 2022. "Algorithms and Influence Artificial Intelligence and Crisis Decision-Making." *International Studies Quarterly* 66 (4): 1–11.

Horowitz, Michael C., Allan C. Stam, & Cali M. Ellis. 2015. *Why Leaders Fight*. Cambridge: Cambridge University Press.

Huntington, Samuel. 1981. *Soldier and the State: The Theory and Politics of Civil-Military Relations.* Cambridge, MA: Belknap Press.

Hurrell, Andrew. 2004. "Legitimacy and the Use of Force: Can the Circle Be Squared?" *Review of International Studies* 31: 15–32.

Hurwitz, Jon & Mark Peffley. 1987. "How Are Foreign Policy Attitudes Structured? A Hierarchical Model." *The American Political Science Review* 81 (4): 1099–120.

Imai, Kosuke, Luke Keele, & Dustin Tingley. 2010. "A General Approach to Causal Mediation Analysis." *Psychological Methods* 15 (4): 309–34.

Imai, Kosuke, Luke Keele, Dustin Tingley, & Teppei Yamamoto. 2011. "Unpacking the Black Box of Causality: Learning about Causal Mechanisms from Experimental and Observational Studies." *American Political Science Review* 105 (4): 765–89.

Irish, John. 2017. "France Turns to Armed Drones in Fight Against Sahel Militants." *Reuters*, September 5, 2017. www.reuters.com/article/us-france-drones-idUSK CN1BG2K2.

Israel, Steve. 2022. Interview with the Authors, March 14, 2022.

Jaeger, David A. & Zahra Siddique. 2018. "Are Drone Strikes Effective in Afghanistan and Pakistan? On the Dynamics of Violence between the United States and the Taliban." *CESifo Economic Studies* 64 (4): 667–97.

Jensen, Benjamin M., Christopher Whyte, & Scott Cuomo. 2022. *Information in War: Military Innovation, Battle Networks, and the Future of Artificial Intelligence.* Washington, DC: Georgetown University Press.

Jervis, Robert. 2017. *Perception and Misperception in International Politics.* Princeton, NJ: Princeton University Press.

Jo, Hyeran. 2015. *Compliant Rebels: Rebel Groups and International Law in World Politics.* Cambridge: Cambridge University Press.

Johnston, Patrick B. & Anoop K. Sarbahi. 2016. "The Impact of US Drone Strikes on Terrorism in Pakistan." *International Studies Quarterly* 60 (2): 203–19.

Jongen, Hortense & Jan Aart Scholte. 2022. "Inequality and Legitimacy in Global Governance: An Empirical Study." *European Journal of International Relations* 28 (3): 1–22.

Kaag, John & Sarah Kreps. 2014. *Drone Warfare.* Oxford: Polity.

Kagan, Robert. 2022. "Power and Weakness." *Hoover Institution Policy Review*, June 1, 2022. www.hoover.org/research/power-and-weakness.

Kaldor, Mary. 2018. *Global Security Cultures.* Cambridge: Cambridge University Press.

Katzenstein, Peter, ed. 1996. *The Cultural of National Security: Norms and Identity in World Politics.* New York: Columbia University Press.

Keating, Vincent C. 2022. "Membership Has Its Privileges: Targeted Killing Norms and the Firewall of International Society." *International Studies Quarterly* 66 (3): 1–12. https://doi.org/10.1093/isq/sqac040.

Keohane, Robert O. 2010. "Social Norms and Agency in World Politics." *Straus Institute Working Paper 07/10.* NYU School of Law, New York. www.law.nyu.edu/sites/default/files/siwp/Keohane.pdf.

Kertzer, Joshua D. 2016. *Resolve in International Politics.* Princeton, NJ: Princeton University Press.

Kertzer, Joshua D. 2017. "Microfoundations in International Relations." *Conflict Management and Peace Science* 34 (1): 81–97.

Kertzer, Joshua D. & Dustin Tingley. 2018. "Political Psychology in International Relations: Beyond the Paradigms." *Annual Review of Political Science* 21: 319–39.

Kertzer, Joshua D. & Thomas Zeitzoff. 2017. "A Bottom-Up Theory of Public Opinion about Foreign Policy." *American Journal of Political Science* 61 (3): 543–58.

Key, V.O. 1961. *Public Opinion and American Democracy*. New York: Alfred A. Knopf.

Khan, Azmat. 2021. Hidden Pentagon Records Reveal Patterns of Failure in Deadly Airstrikes. *The New York Times*, December 8, 2021. www.nytimes.com/interactive/2021/12/18/us/airstrikes-pentagon-records-civilian-deaths.html.

Khan, Lauren. 2022. "How Ukraine is Remaking War: Technological Advancements Are Helping Kyiv Succeed." *Foreign Affairs*, August 29, 2022. www.foreignaffairs.com/ukraine/how-ukraine-remaking-war.

Kibbe, Jennifer D. 2023. "CIA/SOF Convergence and Congressional Oversight." *Intelligence and National Security* 38 (1): 73–89.

Kinder, Donald D. & Cindy D. Kam. 2010. *US Against Them: Ethnocentric Foundations of American Opinion*. Chicago, IL: The University of Chicago Press.

King, Gary, Robert O. Keohane, & Sidney Verba. 1994. *Designing Social Inquiry: Scientific Inference in Qualitative Research*. Princeton, NJ: Princeton University Press.

King, Isabella. 2023. "How France Failed Mali: The End of Operation Barkhane." *Harvard International Review*, January 30, 2023. https://hir.harvard.edu/how-france-failed-mali-the-end-of-operation-barkhane/.

Kneer, Markus & Edouard Machery. 2019. "No Luck for Moral Luck." *Cognition* 182: 331–48.

Kniesner, Thomas J. & W. Kip Viscusi. 2019. *The Value of a Statistical Life*. Oxford: Oxford University Press.

Knuppe, Austin J., Anna O. Pechenkina, & Daniel M. Silverman. 2023. "Civilian Mindsets Toward Peace in Wartime: Evidence from Ukraine." *Working Paper*. https://papers.ssrn.com/sol3/papers.cfm?abstract_id=4459106.

Krasner, Stephen D. 1978. *Defending the National Interest: Raw Materials Investments and U.S. Foreign Policy*. Princeton, NJ: Princeton University Press.

Krasner, Stephen D., ed. 1983. *International Regimes*. Ithaca, NY: Cornell University Press.

Kreps, Sarah. 2011. *Coalitions of Convenience: United States Military Interventions After the Cold War*. Oxford: Oxford University Press.

Kreps, Sarah. 2014. "Flying Under the Radar: A Study of Public Attitudes Towards Unmanned Aerial Vehicles." *Research and Politics* 1 (1): 1–7.

Kreps, Sarah. 2016. *Drones: What Everyone Needs to Know*. Oxford: Oxford University Press.

Kreps, Sarah. 2022. "The U.S. Military Confronts the Darker Side of Drone Strikes." *World Politics Review*, September 16, 2022. www.worldpoliticsreview.com/us-drone-strikes-afghanistan-civilian-casualties/.

Kreps, Sarah, Julie George, Paul Lushenko, & Adi Rao. 2023. "Exploring the Artificial Intelligence 'Trust Paradox': Evidence from a Survey Experiment in the United States." *PLoS ONE* 18 (7): 1–21.

Kreps, Sarah, Douglas Kriner, & Paul Lushenko. 2022. "Calculating the Costs of Covert War: Evidence of Public Attitude Formation from the United States." *Working Paper*. Ithaca, NY: Cornell University.

Kreps, Sarah & Paul Lushenko. 2021. "US Faces Immense Obstacles to Continued Drone War in Afghanistan." *Brookings Tech Stream*, October 19, 2021. www.brookings.edu/techstream/us-faces-immense-obstacles-to-continued-drone-war-in-afghanistan/.

Kreps, Sarah & Sarah Maxey. 2021. "Context Matters: The Transformative Nature of Drones on the Battlefield." In *Technology and International Relations*, edited by

Giampiero Giacomello, Francesco N. Moro, & Marco Valigi, 68–88. Cheltenham: Edward Elgar Publishing.

Kreps, Sarah & Stephen Roblin. 2019. "Treatment Format and External Validity in International Relations Experiments." *International Interactions* 45 (3): 576–94.

Kreps, Sarah & Geoffrey P.R. Wallace. 2016. "International Law, Military Effectiveness, and Public Support for Drone Warfare." *Journal of Peace Research* 5 (6): 830–44.

Kreuzer, Michael. 2016. *Drones and the Future of Air Warfare: The Evolution of Remotely Piloted Aircraft.* London: Routledge.

Kunertova, Dominika. 2023. "The War in Ukraine Shows the Game-Changing Effect of Drones Depends on the Game." *Bulletin of Atomic Scientists* 79 (2): 95–102.

Lake, David A. 2011. "Why 'Isms' Are Evil: Theory, Epistemology, and Academic Sects as Impediments to Understanding and Progress." *International Studies Quarterly* 55 (2): 465–80.

Lebow, Richard N. 2020. *Ethics and International Relations: A Tragic Perspective.* Cambridge: Cambridge University Press.

Lee, Caitlin. 2023. "The Role of Culture in Military Innovation Studies: Lessons Learned from the U.S. Air Force's Adoption of the Predator Drone, 1993–1997." *Journal of Strategic Studies* 46 (1): 115–49.

Leeper, Thomas L., Sara B. Hobolt, & James Tilley. 2020. "Measuring Subgroup Preferences in Conjoint Experiments." *Political Analysis* 28 (2): 207–21.

Lefante, Isabella. 2023. "Regulating the Use of Armed Drones in the Context of Force Short of War." *Boston University International Law Journal* 41 (1): 157–82.

Lewis, Larry & Diane M. Vavrichek. 2016. *Rethinking the Drone War: National Security, Legitimacy, and Civilian Casualties in U.S. Counterterrorism Operations.* Washington, DC: Center for Naval Analysis and Marine Corps University Press.

Lindsay, Jon R. 2020. *Information Technology and Military Power.* Ithaca, NY: Cornell University Press.

Lin-Greenberg, Erik. 2022. "Wargame of Drones: Remotely Piloted Aircraft and Crisis Escalation." *Journal of Conflict Resolution* 66 (10): 1737–65.

Lippmann, Walter. 1922. *Public Opinion.* New York: Harcourt, Brace and Company.

Luft, Aliza. 2020. "Theorizing Moral Cognition: Culture in Action, Situations, and Relationships." *Socius: Sociological Research for a Dynamic World* 6: 1–15.

Lupu, Yonatan & Geoffrey P.R. Wallace. 2022. "Targeting and Public Opinion: An Experimental Analysis in Ukraine." *Journal of Conflict Resolution* 67 (5): 951–78.

Lushenko, Paul. 2022a. "The Moral Legitimacy of Drone Strikes: How the Public Forms Its Judgments." *Texas National Security Review* 6 (1): 1–57.

Lushenko, Paul. 2022b. "The Pentagon's Reckoning with Civilian Casualties Is a Good Start— But It's Only a Start." *Modern War Institute*, September 9, 2022. https://mwi.usma.edu/the-pentagons-reckoning-with-civilian-casualties-is-a-good-start-but-its-only-a-start/.

Lushenko, Paul. 2022c. "U.S. Presidents' Use of Drone Warfare." *Defense & Security Analysis* 38 (1): 31–52.

Lushenko, Paul, Srinjoy Bose, & William Maley, eds. 2022b. *Drones and Global Order: The Implications of Remote Warfare for International Society.* London: Routledge.

Lushenko, Paul, Keith Carter, & Srinjoy Bose. 2023. "Are U.S. Drone Strikes Racially Biased? Evidence of Public Attitude Formation in the United States." *Working Paper.* Ithaca, NY: Cornell University.

Lushenko, Paul & Sarah Kreps. 2022. "What Makes a Drone Strike 'Legitimate' in the Eyes of the Public." *Brookings Tech Stream*, May 5, 2022. www.brookings.

edu/blog/order-from-chaos/2022/05/05/what-makes-a-drone-strike-legitimate-in-the-eyes-of-the-public/.

Lushenko, Paul & Sarah Kreps. 2023a. "Americans Support Exporting Drones to Ukraine—with a Caveat." *The Brookings Institution*, May 25, 2023. www.brookings.edu/blog/order-from-chaos/2023/05/25/americans-support-exporting-drones-to-ukraine-with-a-caveat/.

Lushenko, Paul & Sarah Kreps. 2023b. "Tactical Myths and Perceptions of Reality." *Security Studies*: 1–8.

Lushenko, Paul, Sarah Kreps, & Shyam Raman. 2022a. "A More Just Drone War Is Within Reach: The Case for Tighter Targeting Restrictions." *Foreign Affairs*, January 12, 2022. www.foreignaffairs.com/articles/united-states/2022-01-12/more-just-drone-war-within-reach.

Lushenko, Paul & Shyam Raman. 2022. "CISS Webinar Series 2022 Raising the Standard: Near Certainty and Civilian Harm in Pakistan." *Filmed*, August 5, 2022, in Sydney, New South Wales, Australia: University of Sydney video, 59:28. https://youtu.be/hAdZ-7T4uPc.

Lyall, Jason & Isaiah Wilson III. 2009. "Rage Against the Machines: Explaining Outcomes in Counterinsurgency Wars." *International Organization* 63 (1): 67–106.

MacKinnon, David P., Amanda J. Fairchild, & Matthew S. Fritz. 2007. "Mediation Analysis." *Annual Review of Psychology* 58: 593–614.

Maclean, Ruth. 2020. "Crisis in the Sahel Becoming France's Forever War." *The New York Times*, March 29, 2020. www.nytimes.com/2020/03/29/world/africa/france-sahel-west-africa-.html.

Maclean, Ruth. 2021. "Death of Jihadist Behind Attack on U.S. Soldiers Is Latest Blow for Militants." *The New York Times*, September 16, 2021. www.nytimes.com/2021/09/16/world/africa/isis-sahara.html.

Mahmood, Rafat & Michael Jetter. 2023. "Gone with the Wind: The Consequences of US Drone Strikes in Pakistan." *The Economic Journal* 133 (650): 787–811.

Majnemer, Jacklyn & Gustav Meibauer. 2023. "Names from Nowhere? Fictitious Country Names in Survey Vignettes Affect Experimental Results." *International Studies Quarterly*: 1–10.

Mapes, Jeff. 2013. "Wyden Aids Rand Paul's Senate Filibuster, But Only to a Degree." *The Oregonian*, March 7, 2013. www.oregonlive.com/mapes/2013/03/wyden_joins_rand_pauls_filibus.html.

Matisek, Jahara. 2022. "Libya 2011: Hollow Victory in Low-Cost Air War." In *Air Power in the Age of Primacy: Air Warfare Since the Cold War*, edited by Phil M. Haun, Colin F. Jackson, & Timothy P. Schultz, 177–201. Cambridge: Cambridge University Press.

Maurer, Kevin. 2022. "The Candy Maker, the Cop, and the Fireman Fighting America's Shadow War." *Rolling Stone*, August 28, 2022. www.rollingstone.com/culture/culture-features/niger-west-africa-american-special-forces-report-1396333/.

McDonald, Jack. 2021. "Remote Warfare and the Legitimacy of Military Capabilities." *Defence Studies* 21 (4): 528–44.

Mercer, Jonathan. 1996. *Reputation & International Politics*. Ithaca, NY: Cornell University Press.

Mikhail, John. 2009. "Moral Grammar and Intuitive Jurisprudence: A Formal Model of Unconscious Moral and Legal Knowledge." *Psychology of Learning and Motivation* 50: 27–100.

Miles, Caleb H. 2022. "On the Causal Interpretation of Randomized Interventional Indirect Effects." *Working Paper*. Ithaca, NY: Cornell University. https://arxiv.org/abs/2203.00245.

Miller, Greg. 2012. "White House Approves Broader Yemen Drone Campaign." *Washington Post*, April 25, 2012. www.washingtonpost.com/world/national-security/white-house-approves-broader-yemen-drone-campaign/2012/04/25/gIQA82U6hT_story.html.

Milner, Heather V. & Dustin Tingley. 2013. "The Choice for Multilateralism: Foreign Aid and American Foreign Policy." *The Review of International Organizations* 8 (3): 313–41.

Mittiga, Ross. 2022. "Political Legitimacy, Authoritarianism, and Climate Change." *American Political Science Review* 116 (3): 998–1011.

Moe, Terry M. & William G. Howell. 1999. "The Presidential Power of Unilateral Action." *Journal of Law, Economics, & Organization* 15 (1): 132–79.

Montgomery, Jacob M., Brendan Nyhan, & Michelle Torres. 2018. "How Conditioning on Posttreatment Variables Can Ruin Your Experiment and What to Do About It." *American Journal of Political Science* 62 (3): 760–75.

Morgenthau, Hans. 1948. *Politics Among Nations*. New York: Knopf.

Moyn, Samuel. 2021. *Humane: How the United States Abandoned Peace and Reinvented War*. New York: Farrar, Straus, and Giroux.

Moyn, Samuel. 2022. "'Sweeping' DoD Plan to Mitigate Civilian Harm Merely Humanizes Endless War." *Responsible Statecraft*, August 31, 2022. https://responsiblestatecraft.org/2022/08/31/dod-plan-to-mitigate-civilian-harm-merely-humanizes-endless-war/.

Mueller, Jason C. 2023. "Does the United States Owe Reparations to Somalia?" *Race & Class* 65 (1): 61–82.

Muhammedally, Sahr & Daniel Mahanty. 2022. "The Human Factor: The Enduring Relevance of Protecting Civilians in Future Wars." *Texas National Security Review* 5 (3): 1–21.

Müller, Marcus & Florian Böller. 2022. "The Tug-of-War Over Drone Strike Oversight: Congress and the Politics of Drone Warfare during the Obama Presidency." *Congress & the Presidency* 50 (1): 1–28.

Mummolo, Jonathan. 2018. "Militarization Fails to Enhance Policy Safety or Reduce Crime But May Harm Police Reputation." *PNAS* 115 (37): 9181–86.

Munro, Campbell A.O. 2014. "Mapping the Vertical Battlespace: Towards a Legal Cartography of Aerial Sovereignty." *London Review of International Law* 2 (2): 233–61.

Munro, Campbell A.O. 2015. "The Entangled Sovereignties of Air Police: Mapping the Boundary of the International and the Imperial." *Global Jurist* 15 (2): 117–38.

Muralidharan, Karthik, Mauricio Romero, & Kaspar Wüthrich. 2023. "Factorial Designs, Model Selection, and (Incorrect) Inference in Randomized Experiments." *The Review of Economics and Statistics*: 1–44.

Mutz, Diana C. 2011. *Population-Based Survey Experiments*. Princeton, NJ: Princeton University Press.

Mutz, Diana C. 2022. "The Consequences of Cross-Cutting Networks for Political Participation." *American Journal of Political Science* 46 (4): 838–55.

Nezhat, Omar, Meg Kelly, Alex Horton, & Imogen Piper. 2023. "U.S. Officials Walk Back Claim Drone Strike Killed Senior al-Qaeda Leader." *The Washington Post*, May 18, 2023. www.washingtonpost.com/world/2023/05/18/pentagon-drone-strike-syria-civilian-al-qaeda/.

Nichols, Shaun. 2018. "The Wrong and the Bad." In *Atlas of Moral Psychology*, edited by Kurt Gray & Jesse Graham, 40–45. New York: Guilford Publications.

Obama, Barack. 2013. "Obama's Speech on Drone Policy." *The New York Times*, May 23, 2013. www.nytimes.com/2013/05/24/us/politics/transcript-of-obamas-speech-on-drone-policy.html.

Page, James M. & John Williams. 2022. "Drones, Afghanistan, and Beyond: Towards Analysis and Assessment in Context." *European Journal of International Security* 7: 283–303.

Pan, Christina A., Sahil Yakhmi, Tara P. Iyer, Evan Strasnick, Amy X. Zhang, & Michael S. Berstein. 2022. "Comparing the Perceived Legitimacy of Content Moderation Processes: Contractors, Algorithms, Expert Panels, and Digital Juries." *Proceedings of the ACM on Human-Computer Interaction* 6 (CSCW1), Article No. 82: 1–32.

Pape, Robert A. 1996. *Bombing to Win: Air Power and Coercion in War*. Ithaca, NY: Cornell University Press.

Payne, Andrew. 2020. "Presidents, Politics, and Military Strategy: Electoral Constraints during the Iraq War." *International Security* 44 (3): 163–203.

Pearl, Judea & Dana Mackenzie. 2018. *The Book of Why: The New Science of Cause and Effect*. New York: Basic Books.

Perrin, Cédric, Gilbert Roger, Jean-Marie Bockel, & Raymond Vall. 2022. "Drones d'observation et drones armés: un enjeu de souveraineté." *Senat*, February 4, 2022. www.senat.fr/rap/r16-559/r16-5596.html#toc225.

Phelps, Wayne. 2021. *On Killing Remotely: The Psychology of Killing with Drones*. Boston, MA: Little, Brown and Company.

Pitel, Laura & Raya Jalabi. 2022. "Fast, Cheap, and Deadly: The Budget Drone Changing Global Warfare." *Financial Times*, August 25, 2022. www.ft.com/content/948605a1-cf6c-40ea-b403-9a97d72be2cf.

Plaw, Avery, Matthew S. Fricker, & Brian Glyn Williams. 2011. "Practice Makes Perfect? The Changing Civilian Toll of CIA Drone Strikes in Pakistan." *Perspectives on Terrorism* 5 (5/6): 51–69.

Pong, Beryl. 2022. "The Art of Drone Warfare." *Journal of War & Culture Studies* 15 (4): 377–87.

The Pulitzer Prizes. 2023. "Staff of *The New York Times*, Notably Azmat Khan, Contributing Writer." *The Pulitzer Prizes*, Columbia University. www.pulitzer.org/winners/staff-new-york-times-notably-azmat-khan-contributing-writerpulitpfrica/france-sahel-west-africa.html.

Rae, James D. 2014. *Analyzing the Drone Debates: Targeted Killing, Remote Warfare, and Military Technology*. London: Palgrave Macmillan.

Rafiq, Muhammad. 2011. "Estimating the Value of Statistical Life in Pakistan." *The South Asian Network for Development and Environmental Economics Working Paper 63–11*. Kathmandu, NP: SANDEE. www.sandeeonline.org/uploads/documents/publication/940_PUB_WP_63_Muhammad_Rafiq.pdf.

Rayne, Trevor. 2021. "Turkey's Escalating War on the Kurds." *Revolutionary Communist*, August 27, 2021. www.revolutionarycommunist.org/middle-east/turkey/6328-turkey-s-escalating-war-on-the-kurds.

Recchia, Stefano. 2020. "A Legitimate Sphere of Influence: Understanding France's Turn to Multilateralism in Asia." *Journal of Strategic Studies* 43 (4): 508–33.

Recchia, Stefano & Jonathan Chu. 2021. "Validating Threat: IO Approval and Public Support for Joining Military Counterterrorism Coalitions." *International Studies Quarterly* 65 (4): 919–28.

Reeves, Andrew & Jon C. Rogowski. 2016. "Unilateral Power, Public Opinion, and the Presidency." *The Journal of Politics* 78 (1): 137–51.

Regan, Mitt. 2022. *Drone Strike—Analyzing the Impacts of Targeted Killing*. New York: Palgrave Macmillan.

Regan, Mitt. June 5, 022. Personal communication with the authors.

Renic, Neil. 2020. *Asymmetric Killing: Risk Avoidance, Just War, and the Warrior Ethos*. Oxford: Oxford University Press.

Renic, Neil. 2023a. "Remote Warfare: Drivers, Limits, and Challenges." In *Routledge Handbook of the Future of War*, edited by Artur Gruszczak & Sebastian Kaempf. New York: Routledge.

Renic, Neil. 2023b. "The Limits of Remote Warfare: Aligning Values with Interests." *Just Security*, January 18, 2023. www.justsecurity.org/84807/the-limits-of-remote-warfare-aligning-values-with-interests/.

Reus-Smit, Christian & Duncan Snidal. 2010. "Between Utopia and Reality: The Practical Discourses of International Relations." In *The Oxford Handbook of International Relations*, edited by Christian Reus-Smit & Duncan Snidal, 3–37. Oxford: Oxford University Press.

Richardson, Michael. 2018. "Drone Capitalism." *Transformations* 31: 79–98.

Richardson, Michael. 2020. "Drone Cultures: Encounters with Everyday Militarisms." *Continuum: Journal of Media & Cultural Studies* 34 (6): 858–69.

Richardson, Michael. 2022. "Drone Trauma: Violent Mediation and Remote Warfare." *Media, Culture & Society* 45 (1): 202–11.

Rigterink, Anouk S. 2021. "The Wane of Command: Evidence on Drone Strikes and Control within Terrorist Organizations." *American Political Science Review* 115 (1): 31–50.

Rinehart, Christine S. 2018. *Drones and Targeted Killing in the Middle East and Africa: An Appraisal of American Counterterrorism Policies*. Lanham, MD: Lexington Books.

Rogers, James P. 2020. "What Has Been the Most Significant Development in the History of Weaponry?" *BBC History Magazine*, October 2020.

Rogers, James P. 2023a. *Precision: A History of American Warfare*. Manchester: Manchester University Press.

Rogers, James P. 2023b. "Rethinking Remote Warfare." *International Politics* 60 (4): 1–9.

Rogers, James P. & Delina Goxho. 2022. "Light Footprint-Heavy Destabilising Impact in Niger: Why the Western Understanding of Remote Warfare Needs to Be Reconsidered." *International Politics* 60: 790–817.

Rogers, James P. & Dominika Kunertova. 2022. *The Vulnerabilities of the Drone Age Established Threats and Emerging Issues Out to 2035*. Odense, DK: University of Southern Denmark.

Rosendorf, Ondrej, Michal Smetana, & Marek Vranka. 2022. "Autonomous Weapons an Ethical Judgments: Experimental Evidence on Attitudes Toward the Military Use of 'Military Robots'." *Peace and Conflict: Journal of Peace Psychology* 28 (2): 177–83.

Rossiter, Ash. 2018. "Drone Usage by Militant Groups: Exploring Variation in Adoption." *Defense & Security Analysis* 34 (2): 113–26.

Rossiter, Ash. 2023. "Military Technology and Revolutions in Warfare: Priming the Drone Debate." *Defense & Security Analysis* 39 (2): 253–55.

Rossiter, Ash & Brendon J. Cannon. 2022. "Turkey's Rise as a Drone Power: Trial by Fire." *Defense & Security Analysis* 38 (2): 210–29.

Rowling, Charles M. & Joan M. Blauwkamp. 2021. "Hear No Evil, See No Evil: Motivated Reasoning, Drone Warfare, and the Effects of Message Framing on U.S. Public Opinion." *International Journal of Public Opinion Research* 34 (1): 1–19.

Sagan, Scott D. & Benjamin A. Valentino. 2018. "Revisiting Hiroshima in Iran: What Americans Really Think about Using Nuclear Weapons and Killing Noncombatants." *International Security* 42 (1): 41–79.

Salamé, Ghassan. 2019. "Interview with UN Special Representative for Libya Ghassan Salamé." *Filmed*, September 25, 2019. United Nations Political and Peacebuilding Affairs, 6:26. www.youtube.com/watch?v=IB3jie4i7SI.

Savage, Charlie. 2022. "White House Tightens Rules on Counterterrorism Drone Strikes." *The New York Times*, October 7, 2022. www.nytimes.com/2022/10/07/us/politics/drone-strikes-biden-trump.html.

Schinella, Anthony M. 2019. *Bombs Without Boots: The Limits of Airpower*. Washington, DC: Brookings Institution Press.

Schmidt, Dennis R. & John Williams. 2023. "The Normativity of Global Ordering Practices." *International Studies Quarterly* 67 (2): 1–13.

Schmitt, Eric. 2023. "U.S. Commandos Advise Somalis in Fight Against Qaeda Branch." *The New York Times*, February 27, 2023. www.nytimes.com/2023/02/27/us/politics/somalia-commandos-counterterrorism.html.

Schmitt, Eric, Charlie Savage, & Azmat Khan. 2022. "Austin Orders U.S. Military to Step Up Efforts to Prevent Civilian Harm." *The New York Times*, January 27, 2022. www.nytimes.com/2022/01/27/us/politics/us-airstrikes-rand-report.html.

Schmitt, Oliver. 2018. *Allies that Count: Junior Partners in Coalition Warfare*. Washington, DC: Georgetown University Press.

Schneider, Jacquelyn & Julia Macdonald. 2016. "U.S. Public Support for Drone Strikes." *Center for New American Security*. www.cnas.org/publications/reports/u-s-public-support-for-drone-strikes.

Schwartz, Joshua A., Matthew Fuhrmann, & Michael C. Horowitz. 2022. "Do Armed Drones Counter Terrorism, or Are They Counterproductive? Evidence from Eighteen Countries." *International Studies Quarterly* 66 (3): 1–14.

Sewall, Sarah. 2016. *Chasing Success: Air Force Efforts to Reduce Civilian Harm*. Maxwell Air Force Base, AL: Air University Press.

Shah, Aqil. 2018. "Do U.S. Drone Strikes Cause Blowback?: Evidence from Pakistan and Beyond." *International Security* 42 (2): 47–84.

Shane, Scott. 2016. *Objective Troy: A Terrorist, a President, and the Rise of the Drone*. New York: Tim Duggan Books.

Sheehan, Michael A., Erich Marquardt, & Liam Collins, eds. 2022. *Routledge Handbook of U.S. Counterterrorism and Irregular Warfare Operations*. London: Routledge.

Silverman, Daniel. 2019. "What Shapes Civilian Belies about Violent Events? Experimental Evidence from Pakistan." *Journal of Conflict Resolution* 63 (6): 1460–87.

Silverman, Daniel. 2020. "Too Late to Apologize? Collateral Damage, Post Harm Compensation, and Insurgent Violence in Iraq." *International Organization* 74 (4): 867–68.

Simonsohn, Uris. 2022. "Mediation Analysis Is Counterintuitively Invalid." *Data Colada: Thinking about Evidence, and Vice Versa*, September 26, 2022. http://datacolada.org/103.

Sinnott-Armstrong, Walter. 2008. "Framing Moral Intuitions." In *Moral Psychology, Volume 2, The Cognitive Science of Morality: Intuition and Diversity*, edited by Walter Sinnott-Armstrong, 47–76. Cambridge, MA: The MIT Press.

Slim, Hugo. 2022. *Solferino 21: Warfare, Civilians and Humanitarians in the Twenty-First Century*. London: C. Hurst.

Soyaltin-Colella, Digdem & Tolga Demiryol. 2023. "Unusual Middle Power Activism and Regime Survival: Turkey's Drone Warfare and Its Regime-Booting Effects." *Third World Quarterly* 2023: 1–20.

Stanton, Jessica A. 2016. *Violence and Restraint in Civil War: Civilian Targeting in the Shadow of International Law*. Cambridge: Cambridge University Press.

Staunton, Eglantine. 2020. *France, Humanitarian Intervention, and the Responsibility to Protect*. Manchester: Manchester University Press.

Stein, Janice G. 2017. "The Micro-Foundations of International Relations Theory: Psychology and Behavioral Economics." *International Organization* 71 (1): S249–63.

Sterio, Milena. 2018. "Lethal Use of Drones: When the Executive Is the Judge, Jury, and Executioner." *The Independent Review* 23 (1): 35–50.

Suchman, Mark C. 1995. "Managing Legitimacy: Strategic and Institutional Approaches." *The Academy of Management Review* 20 (3): 571–610.

Suong, Clara H., Scott Desposato, & Erik Gartzke. 2023. "Thinking Generically and Specifically in International Relations Survey Experiments." *Research & Politics* 10 (2): 1–6.

Tago, Atsushi & Maki Ikeda. 2015. "An 'A' for Effort: Experimental Evidence on UN Security Council Engagement and Support for US Military Action in Japan." *British Journal of Political Science* 45 (2): 391–410.

Tallberg, Jonas & Michael Zurn. 2019. "The Legitimacy and Legitimation of International Organizations: Introduction and Framework." *The Review of International Organizations* 14 (4): 581–606.

Ternovski, John & Lilla Orr. 2022. "A Note on Increases in Inattentive Online-Survey Takers Since 2020." *Journal of Quantitative Description: Digital Media* 2: 1–35.

Tirman, John. 2011. *The Deaths of Others: The Fate of Civilians in America's Wars*. Oxford: Oxford University Press.

Titiunik, Rocio & Jasjeet Sekhon. 2012. "When Natural Experiments Are Neither Natural Nor Experiments." *American Political Science Review* 106 (1): 35–57.

Tomz, Michael R. & Jessica L.P. Weeks. 2020. "Human Rights and Public Support for War." *Journal of Politics* 82 (1): 182–94.

Tversky, Amos & Daniel Kahneman. 1981. "The Framing of Decisions and the Psychology of Choice." *Science* 201 (4481): 453–58.

Ullah, Imdad. 2021. *Terrorism and the US Drone Attacks in Pakistan*. New York: Routledge.

University of Maryland. 2021. "Global Terrorism Database." www.start.umd.edu/gtd/.

U.S. Department of Defense. 2022a. "Civilian Harm Mitigation and Response Action Plan (CHMR-AP)." *U.S. Department of Defense*, August 25, 2022. https://media.defense.gov/2022/Aug/25/2003064740/-1/-1/1/CIVILIAN-HARM-MITIGATION-AND-RESPONSE-ACTION-PLAN.PDF.

U.S. Department of Defense. 2022b. "Department of Defense Releases Memorandum on Improving Civilian Harm Mitigation and Response." *U.S. Department of Defense*, January 27, 2022. www.defense.gov/News/Releases/Release/Article/2914764/department-of-defense-releases-memorandum-on-improving-civilian-harm-mitigation/.

Verba, Sidney. 1971. "Cross-national Survey Research: The Problem of Credibility." In *Comparative Methods in Sociology: Essays on Trends and Applications*, edited by Ivan Vallier, 309–56. Berkeley, CA: University of California Press.

Vilmer, Jean-Baptiste J. 2016. "A Perspective on France." *Center for New American Security*. https://drones.cnas.org/wp-content/uploads/2016/05/A-Perspective-on-France-Proliferated-Drones.pdf.

Vilmer, Jean-Baptiste J. 2017. "The French Turn to Armed Drones." *War on the Rocks*, September 22, 2017. https://warontherocks.com/2017/09/the-french-turn-to-armed-drones/.

Vilmer, Jean-Baptiste J. 2021. "Not So Remote Drone Warfare." *International Politics* 60: 1–22.

Viscusi, W. Kip & Clayton J. Masterman. 2017. "Income Elasticities and Global Values of a Statistical Life." *Journal of Benefit-Cost Analysis* 8 (2): 226–50.

Voeten, Erik. 2005. "The Political Origins of the UN Security Council's Ability to Legitimize the Use of Force." *International Organization* 59 (3): 527–57.

Wallace, Geoffrey P.R. 2013. "International Law and Public Attitudes Toward Torture: An Experimental Study." *International Organization* 67 (1): 105–40.

Wallace, Geoffrey P.R. 2019. "Supplying Protection: The United Nations and Public Support for Humanitarian Intervention." *Conflict Management and Peace Science* 36 (3): 248–69.

Watts, Stephen. 2008. "Air War and Restraint: The Role of Public Opinion and Democracy." In *Democracy and Security: Preferences, Norms, and Policy-Making*, edited by Michael Evangelista, Harald Muller, & Niklas Schoernig, 53–72. New York: Routledge.

Weber, Max. 1968. *Economy and Society: Volume One*. Berkeley, CA: University of California Press.

Welsh, Jennifer M. 2015. "The Morality of 'Drone Warfare'." In *Drones and the Future of Armed Conflict: Ethical, Legal, and Strategic Implications*, edited by David Cortright, Rachel Fairhurst, & Kristen Wall, 24–45. Chicago, IL: University of Chicago Press.

Wendt, Alexander. 1992. "Anarchy Is What States Make of It: The Social Construction of Power Politics." *International Organization* 46 (2): 391–425.

Wendt, Alexander. 1999. *Social Theory of International Politics*. Cambridge: Cambridge University Press.

Western, Jon. 2005. *Selling Intervention & War: The Presidency, the Media, and the American Public*. Baltimore, MD: The John Hopkins University Press.

Williams, Brian G. 2011. "Accuracy of the U.S. Drone Campaign: The Views of a Pakistani General." *CTC Sentinel* 4 (3): 9–11.

Witt, Stephen. 2022. "The Turkish Drone that Changed the Nature of Warfare." *The New Yorker*, May 9, 2022. www.newyorker.com/magazine/2022/05/16/the-turkish-drone-that-changed-the-nature-of-warfare.

Yarhi-Milo, Keren. 2018. *Who Fights for Reputation: The Psychology of Leaders in International Conflicts*. Princeton, NJ: Princeton University Press.

Zaller, John. 1992. *The Nature and Origins of Mass Opinion*. Cambridge: Cambridge University Press.

Appendix A

2 x 2 x 2 factorial and between-subject survey experiment manipulating use, constraint, and consequence

8x Treatment Groups: use (strategy vs. tactic) x constraint (multilateral vs. tactical) x consequence (high vs. low), **1x Control Group:** no variation in treatments

Treatment Scenario #1: Country A uses drone warfare as a strategy. This means that Country A's political leaders frequently use drones for targeted killing in support of national military and political objectives, even if doing so erodes the sovereignty of other states. Country A also prefers drone strikes to raids conducted by ground forces to remove terrorists, especially because this better protects its own soldiers. At the same time, **Country A allows its allies to approve its drone strikes.** Prior to conducting a strike, officials from Country A participate in meetings with their allies to verify intelligence on a target, confirm a target's location, and enforce measures to prevent civilian casualties. Country A, using drone warfare as a strategy and participating as a member of a coalition, conducts a drone strike in Country B against a terrorist. **The strike removes the terrorist but results in a civilian casualty.**
Treatment Conditions: Strategic Use, Multilateral Constraint, High Consequence

Treatment Scenario #2: Country A uses drone warfare as a strategy. This means that Country A's political leaders frequently use drones for targeted killing in support of national military and political objectives, even if doing so erodes the sovereignty of other states. Country A also prefers drone strikes to raids conducted by ground forces to remove terrorists, especially because this better protects its own soldiers. At the same time, **Country A allows its allies to approve its drone strikes.** Prior to conducting a strike, officials from Country A participate in meetings with their allies to verify intelligence on a target, confirm a target's location, and enforce measures to prevent civilian casualties. Country A, using drone warfare as a strategy and participating as a member of a coalition, conducts a drone strike in Country B against a terrorist. **The strike removes the terrorist and results in no civilian casualties.**
Treatment Conditions: Strategic Use, Multilateral Constraint, Low Consequence

Treatment Scenario #3: Country A uses drone warfare as a strategy. This means that Country A's political leaders frequently use drones for targeted killing in support of national military and political objectives, even if doing so erodes the sovereignty of other states. Country A also prefers drone strikes to raids conducted by ground forces to remove terrorists, especially because this better protects its own soldiers. At the same time, **Country A prefers to use drone strikes without consulting with or seeking the approval of its allies.** Prior to conducting a strike, officials from Country A convene a domestic-level meeting to verify intelligence on a target, confirm a target's location, and enforce measures to prevent civilian casualties. Country A, using drone warfare as a strategy and acting alone, conducts a drone strike in Country B against a terrorist. **The strike removes the terrorist but results in a civilian casualty.**
Treatment Conditions: Strategic Use, Unilateral Constraint, High Consequence

Treatment Scenario #4: Country A uses drone warfare as a strategy. This means that Country A's political leaders frequently use drones for targeted killing in support of national military and political objectives, even if doing so erodes the sovereignty of other states. Country A also prefers drone strikes to raids conducted by ground forces to remove terrorists, especially because this better protects its own soldiers. At the same time, **Country A prefers to use drone strikes without consulting with or seeking the approval of its allies.** Prior to conducting a strike, officials from Country A convene a domestic-level meeting to verify intelligence on a target, confirm a target's location, and enforce measures to prevent civilian casualties. Country A, using drone warfare as a strategy and acting alone, conducts a drone strike in Country B against a terrorist. **The strike removes the terrorist and results in no civilian casualties.**
Treatment Conditions: Strategic Use, Unilateral Constraint, Low Consequence

Treatment Scenario #5: Country A uses drone warfare as a tactic. This means that Country A's commanders use drone strikes sparingly in support of limited military objectives to achieve limited military objectives. Country A also prefers drone raids conducted by ground forces to drone strikes to remove terrorists, even if this exposes its own soldiers to heightened risk. At the same time, **Country A allows its allies to approve its drone strikes.** Prior to conducting a strike, officials from other states. Country A also prefers raids conducted by ground forces to drone strikes to remove terrorists, even if this exposes its own soldiers to heightened risk. At the same time, **Country A allows its allies to approve its drone strikes.** Prior to conducting a strike, officials from Country A participate in meetings with their allies to verify intelligence on a target, confirm a target's location, and enforce measures to prevent civilian casualties. Country A, using drone warfare as a tactic and participating as a member of a coalition, conducts a drone strike in Country B against a terrorist. **The strike removes the terrorist but results in a civilian casualty.**
Treatment Conditions: Tactical Use, Multilateral Constraint, High Consequence

Treatment Scenario #6: Country A uses drone warfare as a tactic. This means that Country A's commanders use drone strikes sparingly in support of limited military objectives to achieve limited military objectives. Country A also prefers drone raids conducted by ground forces to drone strikes to remove terrorists, even if this exposes its own soldiers to heightened risk. At the same time, **Country A allows its allies to approve its drone strikes.** Prior to conducting a strike, officials from Country A participate in meetings with their allies to verify intelligence on a target, confirm a target's location, and enforce measures to prevent civilian casualties. Country A, using drone warfare as a tactic and participating as a member of a coalition, conducts a drone strike in Country B against a terrorist. **The strike removes the terrorist and results in no civilian casualties.**
Treatment Conditions: Tactical Use, Multilateral Constraint, Low Consequence

Treatment Scenario #7: Country A uses drone warfare as a tactic. This means that Country A's commanders use drone strikes sparingly in support of limited military objectives to achieve limited military objectives. Country A also prefers raids conducted by ground forces to drone strikes to remove terrorists, even if this exposes its own soldiers to heightened risk. At the same time, **Country A prefers to use drone strikes without consulting with or seeking the approval of its allies.** Prior to conducting a strike, officials from Country A convene a domestic-level meeting to verify intelligence on a target, confirm a target's location, and enforce measures to prevent civilian casualties. Given this information, consider the following scenario. Country A, using drone warfare as a tactic and acting alone, conducts a drone strike in Country B against a terrorist. **The strike removes the terrorist but results in a civilian casualty.**
Treatment Conditions: Tactical Use, Unilateral Constraint, High Consequence

Treatment Scenario #8: Country A uses drone warfare as a tactic. This means that Country A's commanders use drone strikes sparingly in support of limited military objectives to achieve limited military objectives. Country A also prefers raids conducted by ground forces to drone strikes to remove terrorists, even if this exposes its own soldiers to heightened risk. At the same time, **Country A prefers to use drone strikes without consulting with or seeking the approval of its allies.** Prior to conducting a strike, officials from Country A convene a domestic-level meeting to verify intelligence on a target, confirm a target's location, and enforce measures to prevent civilian casualties. Given this information, consider the following scenario. Country A, using drone warfare as a tactic and acting alone, conducts a drone strike in Country B against a terrorist. **The strike removes the terrorist and results in no civilian casualties.**
Treatment Conditions: Tactical Use, Unilateral Constraint, Low Consequence

Figure A.1 Survey Experiment Design

Informed Consent Script

I am asking you to participate in an online research study focused on political attitudes toward conflict. This study is led by Paul Lushenko, a Ph.D. student at Cornell University in the Department of Government. The faculty advisor for this study is Professor Sarah Kreps, also a member of Cornell University's Department of Government.

I will ask you to answer several questions based on a hypothetical scenario regarding one country's use of force. The survey should take approximately 10 to 12 minutes to complete. I do not anticipate that you will incur any risks by participating in the research. The insights gathered from this research will help scholars better understand countries' use of force.

All online surveys are anonymous and confidential, and no one besides the primary researcher will have access to the anonymized responses, which are stored and secured by the online Qualtrics survey software. I anticipate that your participation in this online survey presents no greater risk than your everyday use of the Internet. De-identified data from this study may be shared with the research community to advance understanding of political attitudes toward conflict. I will remove any personal information that could identify you before files are shared with other researchers. Despite these measures, I cannot guarantee the anonymity of your personal data.

Your participation in this online survey is voluntary, and you may refuse to participate before the study begins, discontinue at any time, or skip any questions that make you feel uncomfortable, with no adverse impact on your relationship with Cornell University. However, you will only be paid if you complete the entire survey.

If you have any questions, please do not hesitate to contact Paul Lushenko at pal243@cornell.edu. If you have any questions regarding your rights as a subject in this study, you may contact Cornell University's Institutional Review Board for Human Participants at 607–255–5138 or access its website at www.irb.cornell.edu. You may also report your concerns or complaints anonymously through EthicsPoint at www.hotline.cornell.edu or by calling toll-free at 1-866-293-3077. EthicsPoint is an independent organization that serves as a liaison between Cornell University and the person bringing the complaint so that anonymity can be ensured.

Please click the button below to provide consent to participate in this research and proceed with the survey.

Demographic Variables

a. What is your sex?

- Male
- Female
- Other

b. How old are you?

- Less than 18
- 18–24
- 25–34
- 35–44
- 45–54
- 55–65
- Over 65

c. What racial or ethnic group best describes you?

- American Indian and Alaskan Native
- Asian
- Black
- Hispanic
- Native Hawaiian and Other Pacific Islander
- White, Non-Hispanic

d. What is the highest level of education that you have completed?

- Less than high school
- High school (diploma or GED)
- Some college, but no degree
- 2-year college degree
- 4-year college degree
- Advanced or professional degree (MA, MBA, MD, JD, Ph.D., etc.)

Vignettes

Programming Note: A general prompt is provided to all respondents.

The following HYPOTHETICAL SCENARIO describes one country's use of an unmanned aerial vehicle, often referred to as a drone, to conduct a missile strike. I will describe the circumstances and ask how you understand the legitimacy of the strike. In this case, legitimacy is defined by how right or wrong you perceive the strike to be.

Please pay close attention to how Country X understands and uses drone warfare.

a. Do you agree to read the details very carefully and then give your most thoughtful answers?

- Yes
- No

Programming Note: Treatment scenarios are randomized.

Scenario #1 Treatment (Strategic Use, Multilateral Constraint, High Consequence): Country X uses drone warfare as a strategy. This means that Country X's political leaders frequently use drones for targeted killing in support of national military and political objectives, even if doing so erodes the sovereignty of other states. Country X also prefers drone strikes to raids conducted by ground forces to remove terrorists, especially because this better protects its own soldiers. At the same time, **Country X allows its allies to approve its drone strikes.** Prior to conducting a strike, officials from Country X participate in meetings with their allies to verify intelligence on a target, confirm a target's location, and enforce measures to prevent civilian casualties.

Given this information, consider the following scenario: Country X, using drone warfare as a strategy and participating as a member of a coalition, conducts a drone strike in Country Y against a terrorist. **The strike removes the terrorist but results in a civilian casualty.**

Scenario #2 Treatment (Strategic Use, Multilateral Constraint, Low Consequence): Country X uses drone warfare as a strategy. This means that Country X's political leaders frequently use drones for targeted killing in support of national military and political objectives, even if doing so erodes the sovereignty of other states. Country X also prefers drone strikes to raids conducted by ground forces to remove terrorists, especially because this better protects its own soldiers. At the same time, **Country X allows its allies to approve its drone strikes.** Prior to conducting a strike, officials from Country X participate in meetings with their allies to verify intelligence on a target, confirm a target's location, and enforce measures to prevent civilian casualties.

Given this information, consider the following scenario: Country X, using drone warfare as a strategy and participating as a member of a coalition, conducts a drone strike in Country Y against a terrorist. **The strike removes the terrorist and results in no civilian casualties.**

Scenario #3 Treatment (Strategic Use, Unilateral Constraint, High Consequence): Country X uses drone warfare as a strategy. This means that Country X's political leaders frequently use drones for targeted killing in support of national military and political objectives, even if doing so erodes the sovereignty of other states. Country X also prefers drone strikes to raids conducted by ground forces to remove terrorists, especially because this better protects its own soldiers. At the same time, **Country X prefers to use drone strikes without consulting with or seeking the approval of its allies.** Prior to conducting a strike, officials from Country X convene a domestic-level meeting to verify intelligence on a target, confirm a target's location, and enforce measures to prevent civilian casualties.

Given this information, consider the following scenario: Country X, using drone warfare as a strategy and acting alone, conducts a drone strike in Country Y against a terrorist. **The strike removes the terrorist but results in a civilian casualty.**

Scenario #4 Treatment (Strategic Use, Unilateral Constraint, Low Consequence): Country X uses drone warfare as a strategy. This means that Country X's political leaders frequently use drones for targeted killing in support of national military and political objectives, even if doing so erodes the sovereignty of other states. Country X also prefers drone strikes to raids conducted by ground forces to remove terrorists, especially because this better protects its own soldiers. At the same time, **Country X prefers to use drone strikes without consulting with or seeking the approval of its allies**. Prior to conducting a strike, officials from Country X convene a domestic-level meeting to verify intelligence on a target, confirm a target's location, and enforce measures to prevent civilian casualties.

Given this information, consider the following scenario: Country X, using drone warfare as a strategy and acting alone, conducts a drone strike in Country Y against a terrorist. **The strike removes the terrorist and results in no civilian casualties.**

Scenario #5 Treatment (Tactical Use, Multilateral Constraint, High Consequence): Country X uses drone warfare as a tactic. This means that Country X's commanders use drone strikes sparingly to achieve limited military objectives in support of ground forces during a conflict. As such, drone strikes do not erode the sovereignty of other states. Country X also prefers raids conducted by ground forces to drone strikes to remove terrorists, even if this exposes its own soldiers to heightened risk. At the same time, **Country X allows its allies to approve its drone strikes**. Prior to conducting a strike, officials from Country X participate in meetings with their allies to verify intelligence on a target, confirm a target's location, and enforce measures to prevent civilian casualties.

Given this information, consider the following scenario: Country X, using drone warfare as a tactic and participating as a member of a coalition, conducts a drone strike in Country Y against a terrorist. **The strike removes the terrorist but results in a civilian casualty.**

Scenario #6 Treatment (Tactical Use, Multilateral Constraint, Low Consequence): Country X uses drone warfare as a tactic. This means that Country X's commanders use drone strikes sparingly to achieve limited military objectives in support of ground forces during a conflict. As such, drone strikes do not erode the sovereignty of other states. Country X also prefers raids conducted by ground forces to drone strikes to remove terrorists, even if this exposes its own soldiers to heightened risk. At the same time, **Country X allows its allies to approve its drone strikes**. Prior to conducting a strike, officials from Country X participate in meetings with their allies to verify intelligence on a target, confirm a target's location, and enforce measures to prevent civilian casualties.

Given this information, consider the following scenario: Country X, using drone warfare as a tactic and participating as a member of a coalition, conducts a drone strike in Country Y against a terrorist. **The strike removes the terrorist and results in no civilian casualties.**

Scenario #7 Treatment (Tactical Use, Unilateral Constraint, High Consequence): Country X uses drone warfare as a tactic. This means that Country X's commanders use drone strikes sparingly to achieve limited military objectives in support of ground forces during a conflict. As such, drone strikes do not erode the sovereignty of other states. Country X also prefers raids conducted by ground forces to drone strikes to remove terrorists, even if this exposes its own soldiers to heightened risk. At the same time, **Country X prefers to use drone strikes without consulting with or seeking the approval of its allies**. Prior to conducting a strike, officials from Country X convene a domestic-level meeting to verify intelligence on a target, confirm a target's location, and enforce measures to prevent civilian casualties.

Given this information, consider the following scenario: Country X, using drone warfare as a tactic and acting alone, conducts a drone strike in Country Y against a terrorist. **The strike removes the terrorist but results in a civilian casualty**.

Scenario #8 Treatment (Tactical Use, Unilateral Constraint, Low Consequence): Country X uses drone warfare as a tactic. This means that Country X's commanders use drone strikes sparingly to achieve limited military objectives in support of ground forces during a conflict. As such, drone strikes do not erode the sovereignty of other states. Country X also prefers raids conducted by ground forces to drone strikes to remove terrorists, even if this exposes its own soldiers to heightened risk. At the same time, **Country X prefers to use drone strikes without consulting with or seeking the approval of its allies**. Prior to conducting a strike, officials from Country X convene a domestic-level meeting to verify intelligence on a target, confirm a target's location, and enforce measures to prevent civilian casualties.

Given this information, consider the following scenario: Country X, using drone warfare as a tactic and acting alone, conducts a drone strike in Country Y against a terrorist. **The strike removes the terrorist and results in no civilian casualties**.

Control Scenario (No Variation in Use, Constraint, Consequence): Country X uses drone warfare to kill terrorists abroad. Given this information, consider the following scenario: Country X conducts a drone strike in Country Y against a terrorist.

b. In the scenario you just read, did Country X use a drone strike in Country Y against a terrorist?

- Yes
- No
- I don't know

Adjudicating Moral Legitimacy, Moral Responsibility, and Public Support

Programming Note: As per my 2×2×2 factorial and between-subject survey experiment design, respondents will view one of the nine scenarios, ordered randomly, and then answer several questions. I will also track the response time to each of the following questions to determine if—and to what degree—intuitions may clash, resulting in "cognitive wrestling."

a. On a scale of 1 to 10, with 1 representing **"not legitimate"** and 10 representing **"very legitimate,"** how legitimate is Country X's use of the drone strike?

- 1
- 2
- 3
- 4
- 5
- 6
- 7
- 8
- 9
- 10

b. On a scale of 1 to 10, with 1 representing **"not supportive"** and 10 representing **"very supportive,"** how much do you support Country X's use of the drone strike?

- 1
- 2
- 3
- 4
- 5
- 6
- 7
- 8
- 9
- 10

Programming Note: Respondents receiving scenarios 1, 3, 5, and 7 will also receive the following question:

c. On a scale of 1 to 10, with 1 representing **"not responsible"** and 10 representing **"very responsible,"** how responsible is Country X for the civilian casualties that resulted from the drone strike?

- 1

- 2
- 3
- 4
- 5
- 6
- 7
- 8
- 9
- 10

Open-Ended Question

a. What factors did you consider while evaluating the legitimacy of Country X's use of a drone strike in Country Y? Please be as detailed as possible.

Follow-up Questions on Country X

a. On a scale of 1 to 10, with 1 representing "**not representative**" and 10 representing "**very representative**," how representative is Country X of an actual country you have observed operating in the world?

- 1
- 2
- 3
- 4
- 5
- 6
- 7
- 8
- 9
- 10

Programming Note: Respondents selecting 6 or above in 6a will receive the following question:

b. What country do you think Country X best represents? (I will use the fill in the blank option.)

Follow-up Questions on Moral Norms, Moral Rules, and International Law

Programming Note: The following questions (7a–7h) are randomized.

a. What is more legitimate?

- A country that uses drone warfare as a tactic limited to a declared conflict?

- A country that uses drone warfare as a strategy in undeclared conflicts across the globe?

b. What is more legitimate?

- A country that uses drone warfare by itself without the oversight of allies?
- A country that uses drone warfare with the oversight of allies?

c. In assessing the legitimacy of Country X's strike, how important is it for Country X's soldiers to demonstrate physical courage on the battlefield?

- Very Important
- Somewhat Important
- Neutral
- Somewhat Unimportant
- Very Unimportant

d. In assessing the legitimacy of Country X's strike, how important is it for Country X to minimize the risks to its own soldiers on the battlefield?

- Very Important
- Somewhat Important
- Neutral
- Somewhat Unimportant
- Very Unimportant

e. In assessing the legitimacy of Country X's strike, how important is it for Country X to minimize the risks to civilians on the battlefield?

- Very Important
- Somewhat Important
- Neutral
- Somewhat Unimportant
- Very Unimportant

f. In assessing the legitimacy of Country X's strike, how important is it for Country X's legislative branch to authorize the use of force?

- Very Important
- Somewhat Important
- Neutral
- Somewhat Unimportant
- Very Unimportant

g. In assessing the legitimacy of Country X's strike, how important is it for Country X to uphold international law by conducting the strike during a declared conflict?

- Very Important
- Somewhat Important

- Neutral
- Somewhat Unimportant
- Very Unimportant

h. To what extent do you agree with the following statement? "Civilian casualties are never acceptable in war, even if they are unavoidable."

- Strongly Agree
- Agree
- Neither Agree Nor Disagree
- Disagree
- Strongly Disagree

Follow-up Demographic and Dispositional (Conservatism, Ethnocentrism, Foreign Policy Orientation, Internationalism, Religiosity) Variables

Programming Note: The following questions (8a–8o) are randomized.

a. What is your annual income?

- Less than $10,000
- $10,000 to $25,000
- $25,000 to $50,000
- $50,000 to $75,000
- $75,000 to $100,000
- $100,000 or more

(The following questions are designed to assess respondents' conservativism.)

b. Generally speaking, do you usually think of yourself as a . . .?

- Democratic
- Independent
- Republican
- Other
- I am not sure

c. Generally speaking, do you think of yourself as . . .?

- Extremely Liberal
- Liberal
- Slightly Liberal
- Moderate, middle of the road
- Slightly Conservative

- Conservative
- Extremely Conservative
- I am not sure

d. Did you vote in the 2020 U.S. presidential election?

- Yes
- No

Programming Note. Those respondents who answer "Yes" to question 8d will receive the following question:

e. Which candidate did you vote for?

- Joseph Biden
- Donald Trump
- Other

(The following question is designed to assess respondents' ethnocentrism.)

f. To what extent do you agree with the following statement? "When jobs are scarce, employers should give priority to people of this country over immigrants."[1]

- Strongly Agree
- Agree
- Neither Agree Nor Disagree
- Disagree
- Strongly Disagree

(The following questions are designed to assess respondents' foreign policy orientation or preference for the use of force abroad.)

g. To what extent do you agree with the following statement? "The use of military force has a role to play in international affairs."

- Strongly Agree
- Agree
- Neither Agree Nor Disagree
- Disagree
- Strongly Disagree

h. To what extent do you agree with the following statement? "The use of military force only makes problems worse."

- Strongly Agree
- Agree

- Neither Agree Nor Disagree
- Disagree
- Strongly Disagree

(The following questions are designed to assess respondents' internationalism or awareness of global politics.)

i. To what extent do you agree with the following statement? "Great powers such as the U.S. and China need to play an active role in solving problems around the world."[2]

- Strongly Agree
- Agree
- Neither Agree Nor Disagree
- Disagree
- Strongly Disagree

j. To what extent do you agree with the following statement? "International organizations are taking away too much power from countries."

- Strongly Agree
- Agree
- Neither Agree Nor Disagree
- Disagree
- Strongly Disagree

(The following questions are designed to assess respondents' religiosity.)

k. How would you describe your religious affiliation today? (Please select only one.)

- Protestant Christian
- Catholic
- Other Christian
- Jewish
- Muslim
- Buddhist
- Hindu
- Atheist
- No formal religious affiliation
- Other

Programming Note. Those respondents answering "Catholic," "Other Christian," or "Protestant Christian" to question 8k will receive the following question:

l. Are you a "born-again" or evangelical Christian?

- Yes
- No

m. How important is religion in your life?

- Very Important
- Somewhat Important
- Neutral
- Somewhat Unimportant
- Very Unimportant

(The following questions are designed to assess respondents' military status.)

n. Have you ever served or are you currently serving in the U.S. military (including any component—Active Duty, National Guard, Reserves)?

- Yes
- No

o. Has another member of your immediate family (other than you) ever served in the U.S. military (including any component—Active Duty, National Guard, Reserves)?

- Yes
- No

Table A.1 Summary Statistics

		France (N=909)		USA (N=914)		Full Sample (N=1823)	
		N	Pct.	N	Pct.	N	Pct.
Gender	Female	472	51.9	474	51.9	946	51.9
	Male	435	47.9	437	47.8	872	47.8
	Other	2	0.2	3	0.3	5	0.3
Age	18-24	116	12.8	98	10.7	214	11.7
	25-34	123	13.5	141	15.4	264	14.5
	35-44	167	18.4	207	22.6	374	20.5
	45-54	142	15.6	106	11.6	248	13.6
	55-65	251	27.6	168	18.4	419	23.0
	Over 65	110	12.1	194	21.2	304	16.7
Education	2-year collage degree	60	6.6	93	10.2	153	8.4
	4-year collage degree	23	2.5	203	22.2	226	12.4

(Continued)

Table A.1 (Continued)

	High school (diploma or equivalent)	374	41.1	256	28.0	630	34.6
	Less than high school	136	15.0	87	9.5	223	12.2
	Professional degree	184	20.2	114	12.5	298	16.3
	Some collage	132	14.5	161	17.6	293	16.1
Treatment Groups	Control	106	11.7	108	11.8	214	11.7
	Strategic, Multilateral, High	98	10.8	97	10.6	195	10.7
	Strategic, Multilateral, Low	106	11.7	99	10.8	205	11.2
	Strategic, Unilateral, High	102	11.2	104	11.4	206	11.3
	Strategic, Unilateral, Low	106	12.8	108	11.8	224	12.3
	Tectical, Multilateral, High	91	10.0	88	9.6	179	9.8
	Tectical, Multilateral, Low	87	9.6	99	10.8	186	10.2
	Tectical, Unilateral, High	98	10.8	103	11.3	201	11.0
	Tectical, Unilateral, Low	105	11.6	108	11.8	213	11.7

Note: This table shows descriptive statistics for our sample. We dropped respondents who did not identify men or women for ease of analysis. The statistical power for this category would have been too low for any meaningful analysis.

Table A.2 Multivariate Regression Analysis for France

	(1)	(2)	(3)	(4)	Perceptions of Legitimacy (5)	(6)	(7)	(8)	(9)
SMH	-0.14 (-0.81, 0.53)	-0.13 (-0.78, 0.53)	-0.13 (-0.78, 0.53)	-0.06 (-0.70, 0.58)	0.04 (-0.59, 0.66)	0.03 (-0.59, 0.66)	0.06 (-0.57, 0.69)	0.05 (-0.58, 0.67)	0.04 (-0.59, 0.66)
SML	0.45 (-0.20, 1.11)	0.47 (-0.18, 1.11)	0.47 (-0.17, 1.11)	0.48 (-0.14, 1.11)	0.54* (-0.08, 1.15)	0.54* (-0.07, 1.16)	0.55* (-0.06, 1.16)	0.55* (-0.07, 1.16)	0.54* (-0.07, 1.16)
SLH	-0.72** (-1.38, -0.05)	-0.57* (-1.22, 0.08)	-0.57* (-1.22, 0.09)	-0.50 (-1.14, 0.13)	-0.50 (-1.12, 0.12)	-0.51 (-1.13, 0.11)	-0.49 (-1.11, 0.13)	-0.50 (-1.12, 0.12)	-0.52* (-1.14, 0.10)
SUL	-0.06 (-0.70, 0.58)	0.06 (-0.57, 0.69)	0.06 (-0.57, 0.69)	0.10 (-0.52, 0.71)	0.19 (-0.41, 0.79)	0.19 (-0.41* 0.79)	0.20 (-0.40, 0.80)	0.20 (-0.40, 0.80)	0.19 (-0.41, 0.79)
TMH	-0.11 (-0.80, 0.57)	-0.11 (-0.78, 0.56)	-0.10 (-0.77, 0.57)	0.08 (-0.58, 0.74)	0.09 (-0.55, 0.73)	0.09 (-0.55, 0.73)	0.11 (-0.53, 0.75)	0.10 (-0.54] 0.74)	0.08 (-0.56, 0.72)
TML	0.61* (-0.08, 1.30)	0.73** (0.05, 1.41)	0.73** (0.06, 1.41)	0.73** (0.07, 1.39)	0.67** (0.02, 1.31)	0.66** (0.01, 1.30)	0.65** (-0.0002, 1.29)	0.64** (-0.004, 1.29)	0.62* (-0.02, 1.27)
TLH	-0.53 (-1.21, 0.14)	-0.38 (-1.04, 0.28)	-0.38 (-1.04, 0.28)	-0.40 (-1.04, 0.25)	-0.30 (-0.93, 0.33)	-0.29 (-0.92, 0.34)	-0.27 (-0.90, 0.36)	-0.26 (-0.89, 0.37)	-0.28 (-0.91, 0.35)
TUL	-0.24 (-0.99, 0.42)	-0.09 (-0.74, 0.55)	-0.09 (-0.73, 0.56)	-0.02 (-0.66, 0.61)	0.09 (-0.53, 0.71)	0.09 (-0.53, 0.70)	0.0 (-0.52, 0.71)	0.10 (-0.51, 0.72)	0.09 (-0.53, 0.70)
Age		0.19*** (0.09, 0.29)	0.19*** (0.09, 0.29)	0.19*** (0.09, 0.29)	0.22*** (0.12, 0.32)	0.23*** (0.13, 0.32)	0.21*** (0.12, 0.31)	0.21*** (0.11, 0.31)	0.21*** (0.11, 0.31)
Sex		-0.66*** (-0.98, -0.34)	-0.67*** (-0.99, -0.35)	-0.58*** (-0.89, -0.26)	-0.53*** (-0.83, -0.22)	-0.50*** (-0.81, -0.19)	-0.50*** (-0.81, -0.19)	-0.44*** (-0.77, -0.11)	-0.45*** (-0.79, -0.12)
Education			-0.002 (-0.10, 0.10)	0.04 (-0.06, 0.13)	0.04 (-0.05, 0.14)	0.05 (-0.05, 0.14)	0.05 (-0.05, 0.14)	0.05 (-0.05, 0.14)	0.05 (-0.05, 0.14)
Race		-0.07 (-0.23, 0.09)	-0.07 (-0.23, 0.09)	-0.03 (-0.19, 0.13)	-0.04 (-0.19, 0.11)	-0.04 (-0.19, 0.11)	-0.03 (-0.19, 0.12)	-0.03 (-0.18, 0.13)	-0.02 (-0.18, 0.13)
Income	0.17** (0.01, 0.34)	0.17** (0.01, 0.34)	0.17** (0.01, 0.34)	0.17** (0.01, 0.33)	0.14** (-0.02, 0.29)	0.13 (-0.03, 0.29)	0.12 (-0.04, 0.28)	0.11 (-0.05, 0.27)	0.51* (-0.08, 1.10)

(*Continued*)

Table A.2 (Continued)

	(1)	(2)	(3)	(4)	(5)	(6)	(7)	(8)	(9)
Political Ideology			0.01 (-0.06, 0.08)	-0.01 (-0.08, 0.06)	-0.01 (-0.08, 0.06)	-0.004 (-0.07, 0.06)	-0.01 (-0.07, 0.06)	-0.004 (-0.07, 0.06)	-0.0003 (-0.07, 0.07)
Ethnocentrism			0.45*** (0.32, 0.59)	0.30*** (0.16, 0.44)	0.28*** (0.14, 0.43)	0.29*** (0.14, 0.43)	0.28*** (0.14, 0.43)	0.28*** (0.13, 0.42)	0.27*** (0.13, 0.42)
Foreign Policy				0.66*** (0.43, 0.77)	0.60*** (0.43, 0.77)	0.57*** (0.40, 0.74)	0.58*** (0.41, 0.75)	0.58*** (0.40, 0.75)	0.84*** (0.43, 1.26)
Global Politics					0.13*** (-0.02, 0.29)	0.14*** (-0.01, 0.30)	0.14*** (-0.01, 0.30)	0.14*** (-0.01, 0.38)	0.13* (-0.02, 0.29)
Religiosity						-0.08 (-0.20, 0.03)	-0.08 (-0.20, 0.03)	-0.05 (-0.02, 0.02)	-0.08 (-0.20, 0.03)
Military Service								-0.19 (-0.57, 0.18)	-0.18 (-0.56, 0.19)
Income:FP Orientation									-0.11 (-0.26, 0.05)
Constant	6.90*** (6.44, 7.37)	6.62*** (5.65, 7.59)	6.55*** (5.52, 7.59)	4.64*** (3.47, 5.80)	2.85*** (1.60, 4.09)	2.45*** (1.12, 3.77)	2.66*** (1.30, 4.01)	2.97*** (1.49, 4.45)	2.02** (0.02, 4.02)
N	907	907	907	907	907	907	907	907	907
Adjusted R2	0.02	0.06	0.10	0.15	0.15	0.15	0.15	0.15	0.15
F Statistic	2.89***	5.64***	5.24***	7.89***	10.83***	10.38***	9.94***	9.47***	9.10***
Notes:							***Significant at the 1 percent level.		
							**Significant at the 5 percent level.		
							*Significant at the 10 percent level.		

Note: The dependent variable is respondents' perceptions of the legitimacy of the strike. The coefficients for the eight treatment groups represent the legitimacy outcomes relative to the control group, which is the baseline and not shown. The coefficients for the demographic and ideological factors represent the legitimacy outcomes given a one-unit increase in age. For instance, a one-unit increase in age, indicating a move from younger to older, increases the legitimacy outcome by 0.19 points in Model 2, which is statistically significant at the $p = 0.01$ level. All models, which add additional demographic variables, are standard OLS regressions. 95% confidence intervals are presented in parentheses. "S" refers to "strategic." "T" refers to "tactical." "M" refers to "multilateral." "U" refers to "unilateral." "H" refers to a "high"—or one—civilian casualty outcome. "L" refers to a "low"—or no—civilian casualty outcome.

Table A.3 Multivariate Regression Analysis for the United States

	(1)	(2)	(3)	(4)	(5)	(6)	(7)	(8)	(9)
					Perceptions of Legitimacy				
SMH	-0.13 (-0.79, 0.53)	-0.18 (-0.82, 0.47)	-0.17 (-0.82, 0.47)	-0.14 (-0.78, 0.49)	-0.17 (-0.80, 0.45)	-0.16 (-0.79, 0.46)	-0.17 (-0.80, 0.45)	-0.19 (-0.81, 0.44)	-0.16 (-0.79, 0.46)
SML	0.63* (-0.03, 1.29)	0.71** (0.06, 1.15)	0.73** (0.08, 1.37)	0.70** (0.06, 1.33)	0.67** (0.04, 1.29)	0.66** (0.04, 1.29)	0.65** (0.03, 1.28)	0.66** (0.03, 1.28)	0.67** (0.05, 1.29)
SUH	-0.76** (-1.41, -0.11)	-0.70** (-1.34, -0.07)	-0.70** (-1.34, -0.06)	-0.61* (-1.24, 0.01)	-0.61* (-1.23, 0.01)	-0.63** (-1.25, -0.02)	-0.63** (-1.25, -0.01)	-0.63** (-1.25, -0.02)	-0.62** (-1.23, -0.003)
SUL	0.74** (0.10, 1.38)	0.65** (0.02, 1.28)	0.65** (0.02, 1.28)	0.56* (-0.06, 1.18)	0.55* (-0.06, 1.16)	0.54* (-0.07, 1.15)	0.52* (-0.09, 1.13)	0.51* (-0.10, 1.12)	0.52* (-0.09, 1.12)
TMH	-0.25, (-0.43, 0.93)	0.30 (-0.36, 0.97)	0.31 (-0.36, 0.97)	0.33 (-0.32, 0.98)	0.22 (-0.42, 0.86)	0.23 (-0.41, 0.87)	0.21 (-0.43, 0.85)	0.22 (-0.41, 0.86)	0.22 (-0.42, 0.86)
TML	0.30 (-0.36, 0.96)	0.32 (-0.32, 0.96)	0.33 (-0.31, 0.98)	0.32 (-0.32, 0.95)	0.41 (-0.21, 1.03)	0.41 (-0.21, 1.03)	0.38 (-0.24, 1.00)	0.39 (-0.24, 1.01)	0.39 (-0.24, 1.01)
TUH	-0.45 (-1.10, 0.20)	-0.45 (-1.08, 0.19)	-0.44 (-1.07, 0.20)	-0.43 (-1.06, 0.20)	-0.49 (-1.11, 0.12)	-0.49 (-1.11, 0.12)	-0.51 (-1.13, 0.10)	-0.50 (-1.12, 0.12)	-0.53* (-1.15, 0.08)
TUL	0.37 (-0.27, 1.01)	0.38 (-0.25, 1.00)	0.39 (-0.23, 1.02)	0.33 (-0.28, 0.95)	0.38 (-0.23, 0.99)	0.38 (-0.23, 0.99)	0.38 (-0.21, 0.99)	0.37 (-0.24, 0.98)	0.36 (-0.25, 0.96)
Age		0.16*** (0.06, 0.26)	0.16*** (0.05, 0.26)	0.14*** (0.05, 0.24)	0.14*** (0.04, 0.24)	0.14*** (0.04, 0.24)	0.13*** (0.04, 0.23)	0.14*** (0.04, 0.24)	0.14*** (0.04, 0.24)
Sex		-0.33* (-0.69, 0.02)	-0.33* (-0.69, 0.03)	-0.29 (-0.64, 0.06)	-0.27 (-0.62, 0.07)	-0.26 (-0.60, 0.09)	-0.28 (-0.62, 0.07)	-0.24 (-0.59, 0.11)	-0.27 (-0.62, 0.08)
Education		0.01 (-0.11, 0.14)	0.01 (-0.12, 0.13)	0.01 (-0.12, 0.13)	0.01 (-0.11. 0.13)	0.01 (-0.11, 0.14)	0.01 (-0.11, 0.14)	0.01 (-0.11, 0.13)	0.002 (-0.12, 0.12)
Race	0.14** (0.02, 0.27)	0.14** (0.02, 0.27)	0.15** (0.02, 0.28)	0.12* (-0.01, 0.24)	0.11* (-0.02, 0.24)	0.10* (-0.02, 0.23)	0.11* (-0.01, 0.24)	0.12* (-0.01, 0.24)	0.12* (-0.0003, 0.25)
Income	0.20*** (0.08, 0.32)	0.20*** (0.08, 0.32)	0.20*** (0.08, 0.32)	0.19*** (0.07, 0.31)	0.17*** (0.05, 0.28)	0.16*** (0.04, 0.28)	0.15*** (0.04, 0.27)	0.15** (0.03, 0.27)	0.63*** (0.17, 1.09)

(Continued)

Table A.3 (Continued)

	(1)	(2)	(3)	(4)	(5)	(6)	(7)	(8)	(9)
Conservatism			-0.05 (-0.20, 0.10)	-0.09 (-0.23, 0.06)	-0.04 (-0.19, 0.10)	-0.03 (-0.17, 0.12)	-0.03 (-0.18, 0.11)	-0.03 (-0.18, 0.11)	-0.03 (-0.17, 0.12)
Ethnocentricism				0.37*** (0.24, 0.51)	0.27*** (0.14, 0.41)	0.26*** (0.13, 0.40)	0.25*** (0.11, 0.39)	0.24*** (0.11, 0.38)	0.24*** (0.10, 0.38)
Foreign Policy					0.53*** (0.36, 0.71)	0.49*** (0.31, 0.67)	0.48*** (0.30, 0.66)	0.47*** (0.29, 0.65)	0.89*** (0.46, 1.31)
Global Politics						0.12 (-0.03, 0.27)	0.12 (-0.03, 0.27)	0.11 (-0.04, 0.27)	0.10 (-0.06, 0.25)
Religiosity							0.07 (-0.03, 0.18)	0.07 (-0.04, 0.18)	0.07 (-0.04, 0.18)
Military Service								-0.37 (-0.83, 0.10)	-0.42* (-0.89, 0.05)
Income:FP Orientation									-0.12** (-0.23, -0.01)
Constant	6.86*** (6.41, 7.32)	5.12*** (4.07, 6.18)	5.21*** (4.13, 6.30)	4.02*** (2.87, 5.17)	2.41*** (1.16, 3.66)	2.08*** (0.76, 3.39)	1.99*** (0.67, 3.31)	2.79*** (1.10, 4.30)	1.28 (-0.80, 3.36)
N	911	911	911	911	911	911	911	911	911
Adjusted R2	0.03	0.08	0.08	0.11	0.14	0.14	0.14	0.14	0.15
F Statistic	4.32***	7.07***	6.59***	8.35***	10.34***	9.89***	9.45***	9.09***	8.89***
Notes:								***Significant at the 1 percent level.	
								**Significant at the 5 percent level.	
								*Significant at the 10 percent level.	

Note: The dependent variable is respondents' perceptions of the legitimacy of the strike. The coefficients for the eight treatment groups represent the legitimacy outcomes relative to the control group, which is the baseline and not shown. The coefficients for the demographic and ideological factors represent the legitimacy outcomes given a one-unit increase in these factors. For instance, a one-unit increase in age, which is statistically significant at the $p = 0.01$ level. All models, which add additional demographic variables, are standard OLS regressions. 95% confidence intervals are presented in parentheses. "S" refers to "strategic." "T" refers to "tactical." "M" refers to "multilateral." "U" refers to "unilateral." "H" refers to a "high"—or one—civilian casualty outcome. "L" refers to a "low"—or no—civilian casualty outcome.

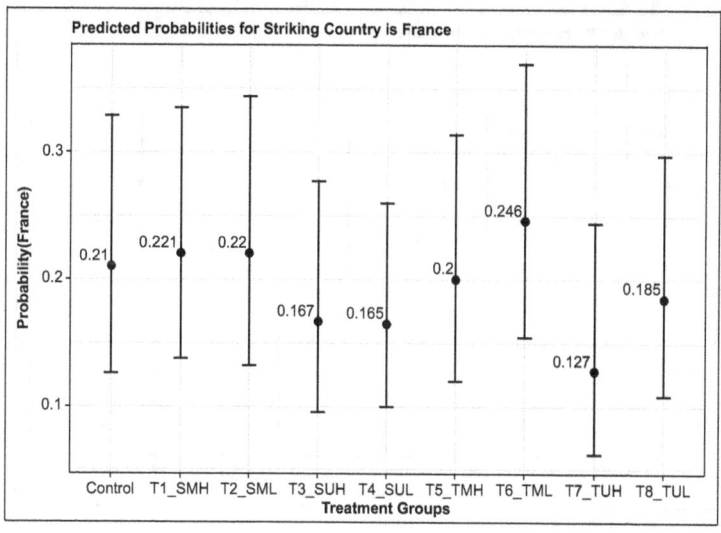

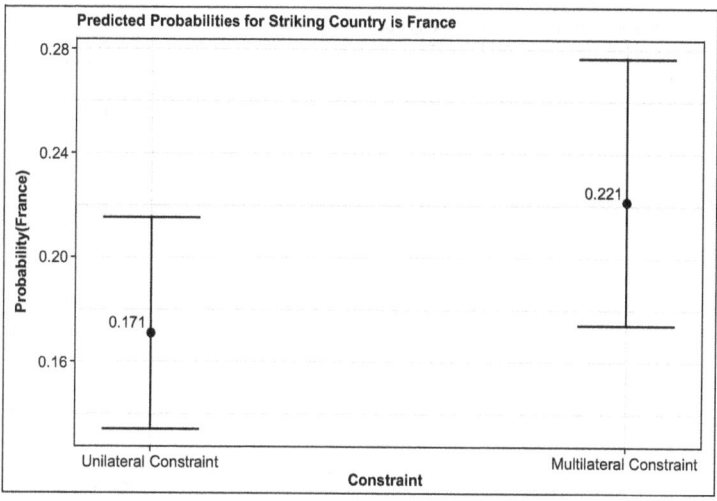

Figure A.2 Logit Models for France

Note: Marginal effects obtained from a logit regression model. Vertical bars present 95% confidence about each point estimate. This plots the change in predicted probability produced by changing the indicator variable from 0 to 1. "S" refers to "strategic." "T" refers to "tactical." "M" refers to "multilateral." "U" refers to "unilateral." "H" refers to a "high"—or one—civilian casualty outcome. "L" refers to a "low"—or no—civilian casualty outcome.

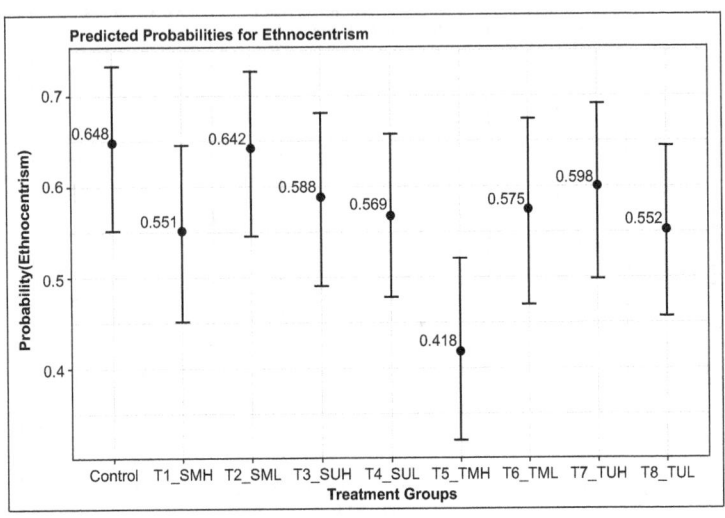

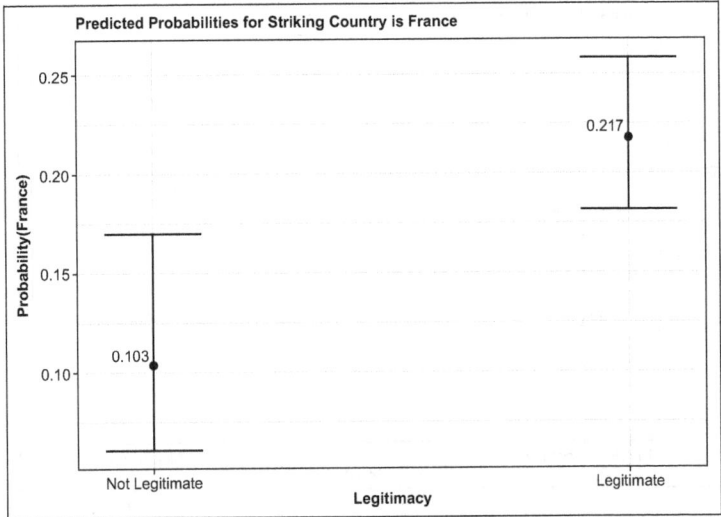

Figure A.2 (Continued)

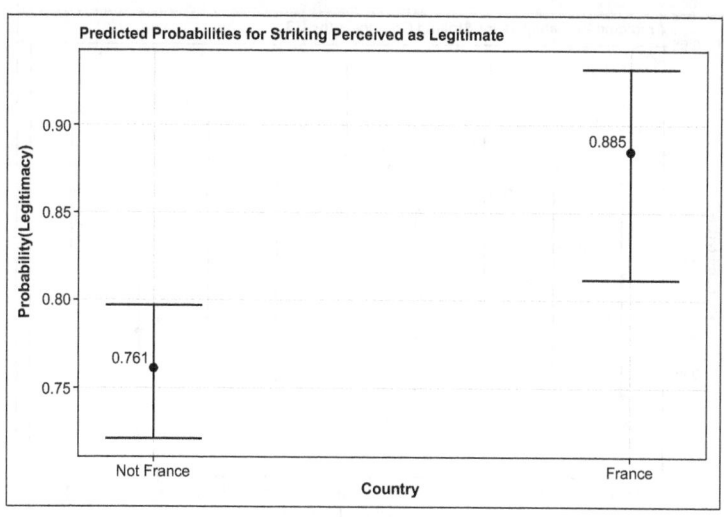

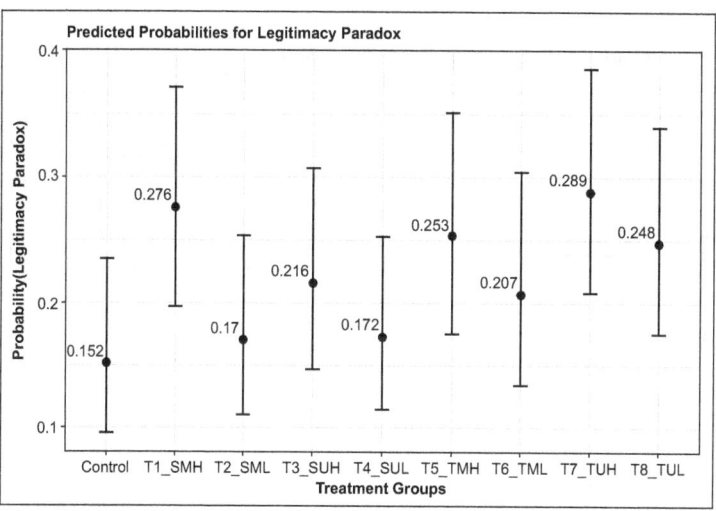

Figure A.2 (Continued)

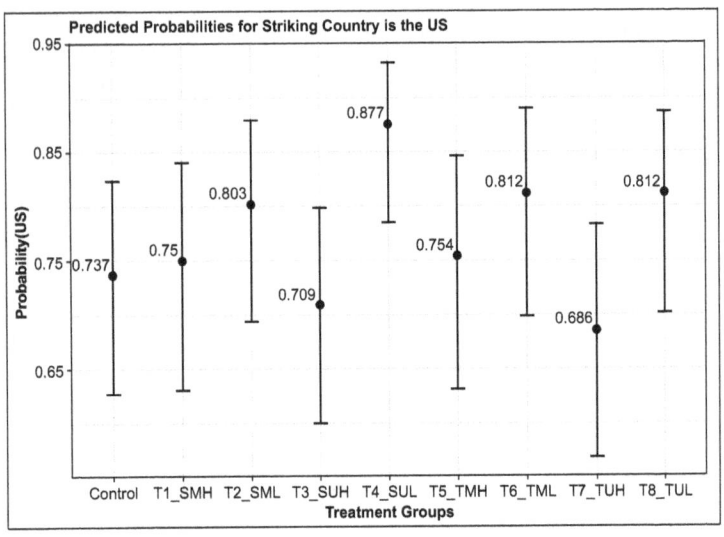

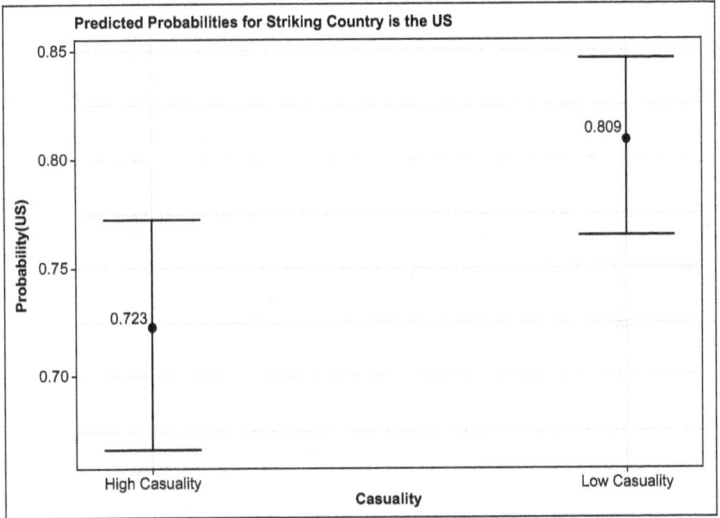

Figure A.3 Logit Models for the United States

Note: Marginal effects obtained from a logit regression model. Vertical bars present 95% confidence about each point estimate. This plots the change in predicted probability produced by changing the indicator variable from 0 to 1. "S" refers to "strategic." "T" refers to "tactical." "M" refers to "multilateral." "U" refers to "unilateral." "H" refers to a "high"—or one—civilian casualty outcome. "L" refers to a "low"—or no—civilian casualty outcome.

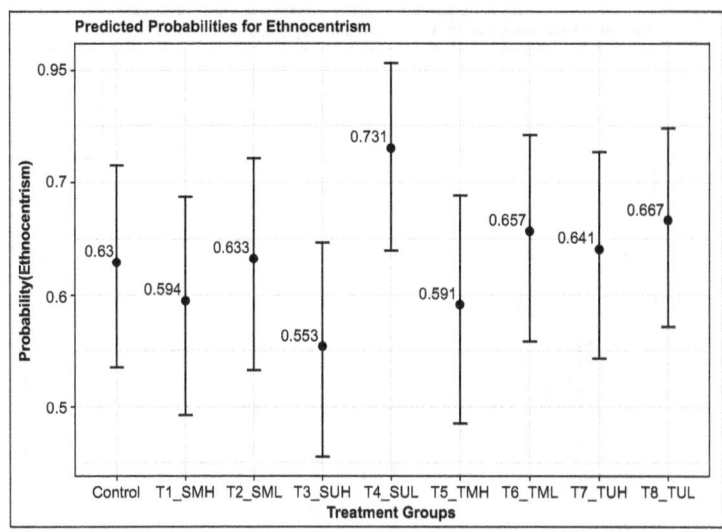

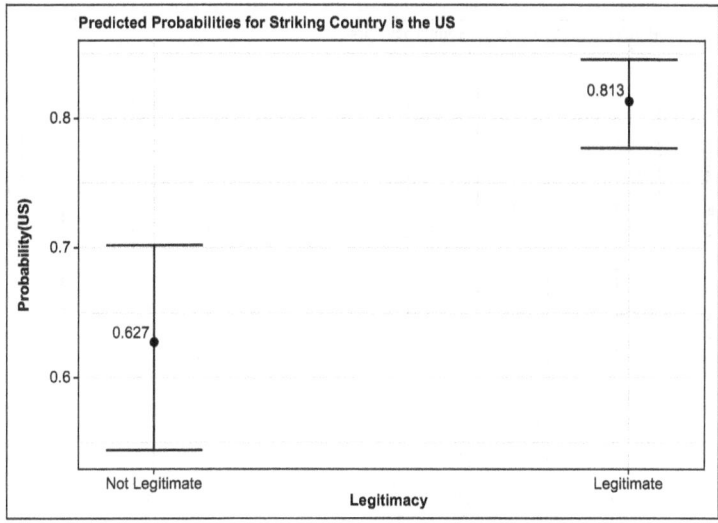

Figure A.3 (Continued)

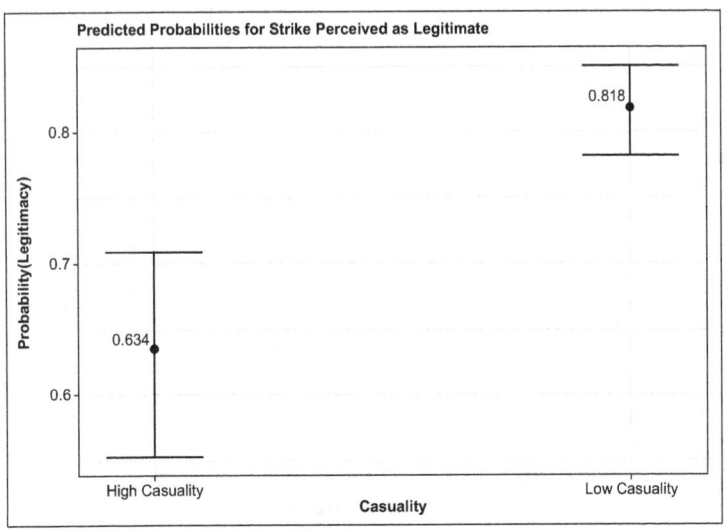

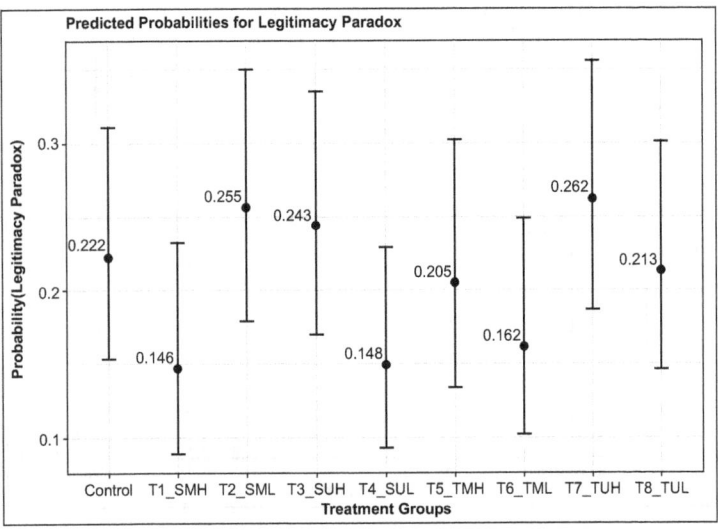

Figure A.3 (Continued)

Table A.4 Analysis of Mediators, the "French Model," and Perceptions of Legitimacy

Mediator	Average Causal Mediation Effect	% of Total Effect Mediated
Civilian Causalities	.074 (-.007, .190)*	8
Domestic Authorization	.035 (-.092, .170)	4
Great Power Responsibilities	.076 (-.014, .190)*	8
International Approval	-.006 (-.117, .110)	-.7†
Support to the Use of Force	.201 (.038, .388)**	22

Note: 95% confidence intervals are in parentheses for each mediator's average causal mediation effect. *** denotes $p < 0.001$; ** $p < 0.01$; and, * $p < 0.05$. †Some values are negative because the effect of the mediator acts in the opposite direction of the effect of the treatment (e.g., "French Model").

Notes

1 Question adopted from "2017–2021 World Values Survey Wave 7 Master Survey Questionnaire," available from www.worldvaluessurvey.org.

2 Question adopted from Hermann et al. (2001).

Appendix B

Table B.1 Suicide Attacks in Pakistan From 2009 to 2018

Group	Overall, N = 631	2009, N = 73	2010, N = 43	2011, N = 58	2012, N = 100	2013, N = 130	2014, N = 89	2015, N = 39	2016, N = 35	2017, N = 37	2018, N = 27
Attack, n (%)											
Individual	258 (41)	30 (41)	24 (56)	25 (43)	29 (29)	45 (35)	29 (33)	22 (56)	25 (71)	17 (46)	12 (44)
Vehicle	373 (59)	43 (59)	19 (44)	33 (57)	71 (71)	85 (65)	60 (67)	17 (44)	10 (29)	20 (54)	15 (56)

Note: Data from the Global Terrorism Database compiled by the University of Maryland. The table shows the distribution—in terms of absolute number and percent of the total—of individual and vehicle suicide attacks in Pakistan by year from 2009 to 2018.

Appendix C

2 x 2 factorial and between-subject survey experiment manipulating the targeting country and UN approval.

Four Treatment Groups: country (U.S. vs. France) x UN approval (has vs. has not); **One Control Group:** no variation in country and UN approval.

Treatment Scenarios #1-4:

- A terrorist group has enjoyed sanctuary in a fragile country in a volatile region of the world.
- The terrorist group has planned to attack a western country with the intent to kill civilians.
- Given the intelligence, the United Nations **[has/has not]** authorized the use of force including armed drones to prevent the attack.
- The **[U.S./France]** decides to use a drone strike to kill the terrorists responsible for planning the attack. Although the strike removes the terrorists, the terror group's leaders claim the operation also resulted in collateral damage, including civilian casualties. **[American/French]** officials adamantly deny this claim.

Treatment Conditions: *Country* and *UN approval.*

Control Scenario:

- A terrorist group has enjoyed sanctuary in a fragile country in a volatile region of the world.
- The terrorist group has planned to attack a western country with the intent to kill civilians.
- Given the intelligence, an **unnamed country** decides to use of force including armed drones to prevent the attack.
- Although the strike removes the terrorists, the terror group's leaders claim the operation also resulted in collateral damage, including civilian casualties. Officials from **unnamed country** adamantly deny this claim.

Control Conditions: No variation in *Country* and *UN approval.*

Figure C.1 Survey Experiment Design

Informed Consent Script

I am asking you to participate in an online research study focused on political attitudes toward conflict. This study is led by Paul Lushenko, a Ph.D. student at Cornell University in the Department of Government. The faculty advisor for this study is Professor Sarah Kreps, also a member of Cornell University's Department of Government.

I will ask you to answer several questions based on a hypothetical scenario regarding one country's use of force. The survey should take approximately 10 to 12 minutes to complete. I do not anticipate that you will incur any risks by participating in the research. The insights gathered from this research will help scholars better understand countries' use of force.

All online surveys are anonymous and confidential, and no one besides the primary researcher will have access to the anonymized responses, which are stored and secured by the online Qualtrics survey software. I anticipate that your participation in this online survey presents no greater risk than your everyday use of the Internet. De-identified data from this study may be shared with the research community to advance understanding of political attitudes toward conflict. I will remove any personal information that could identify you before files are shared with other researchers. Despite these measures, I cannot guarantee the anonymity of your personal data.

Your participation in this online survey is voluntary, and you may refuse to participate before the study begins, discontinue at any time, or skip any questions that make you feel uncomfortable, with no adverse impact on your relationship with Cornell University. However, you will only be paid if you complete the entire survey.

If you have any questions, please do not hesitate to contact Paul Lushenko at pal243@cornell.edu. If you have any questions regarding your rights as a subject in this study, you may contact Cornell University's Institutional Review Board for Human Participants at 607–255–5138 or access its website at www.irb.cornell.edu. You may also report your concerns or complaints anonymously through EthicsPoint at www.hotline.cornell.edu or by calling toll-free at 1-866-293-3077. EthicsPoint is an independent organization that serves as a liaison between Cornell University and the person bringing the complaint so that anonymity can be ensured.

Please click the button below to provide consent to participate in this research and proceed with the survey.

Demographic Variables

a. What is your sex?

- Male
- Female
- Other

b. How old are you?

- Less than 18
- 18–24
- 25–34
- 35–44
- 45–54
- 55–65
- Over 65

c. What racial or ethnic group best describes you?

- American Indian and Alaskan Native
- Asian
- Black
- Hispanic
- Native Hawaiian and Other Pacific Islander
- White, Non-Hispanic

d. What is the highest level of education that you have completed?

- Less than high school
- High school
- Some college, but no degree
- 2-year college degree
- 4-year college degree
- Advanced or professional degree (MA, MBA, MD, JD, Ph.D., etc.)

Vignettes

Programming Note: A general prompt is provided to all respondents.

I will now provide you with a HYPOTHETICAL SCENARIO that describes one country's use of an armed unmanned aerial vehicle, often referred to as a drone, to conduct a missile strike.

Please read the details very carefully.

a. Do you agree to read the details very carefully and then give your most thoughtful answers?

- Yes
- No

Programming Note: Order is randomized.
Here are the circumstances:

- A terrorist group has enjoyed sanctuary in a fragile country in a volatile region of the world.

- The terrorist group has planned to attack a western country with the intent of killing civilians.
- Given the intelligence, the United Nations has/has not authorized the use of force, including armed drones, to prevent the attack.
- The United States/France decides to use a drone strike to kill the terrorists responsible for planning the attack. Although the strike removes the terrorists, the terror group's leaders claim the operation also resulted in collateral damage, including civilian casualties. American/French officials adamantly deny this claim.

Adjudicating the Public's Support and Perception of Legitimacy

Programming Note: As per the 2×2 factorial and between-subject survey experiment design, respondents will view one of the five scenarios and then answer several questions.

Programming Note: The order of 4a and 4b is randomized.

a. Do you favor or oppose the use of drone strikes against terrorists?

- Favor Strongly
- Favor Somewhat
- Neither Favor Nor Oppose
- Oppose Somewhat
- Oppose Strongly

b. In your estimation, how legitimate or rightful was the drone strike?

- Very Legitimate
- Somewhat Legitimate
- Neither Legitimate Nor Illegitimate
- Somewhat Illegitimate
- Very Illegitimate

Open-Ended Questions

a. What factors did you consider while evaluating your support for the drone strike? Please be as detailed as possible.

Follow-up Questions on Signal of Merit, Legality, Morality, and Burden-Sharing

Just to review:

- A terrorist group has enjoyed sanctuary in a fragile country in a volatile region of the world.

- The terrorist group has planned to attack a western country with the intent of killing civilians.
- Given the intelligence, the United Nations has/has not authorized the use of force, including armed drones, to prevent the attack.
- The United States/France decides to use a drone strike to kill the terrorists responsible for planning the attack. Although the strike removes the terrorists, the terror group's leaders claim the operation also resulted in collateral damage, including civilian casualties. American/French officials adamantly deny this claim.

	Almost no chance	*25% chance*	*50–50 chance*	*75% chance*	*Nearly 100% certain*
If the United States/ France does not conduct the drone strike, will the terrorist group **continue to enjoy sanctuary**?	❏	❏	❏	❏	❏
If the United States/ France does not conduct the drone strike, will the terrorist group **launch an attack**?	❏	❏	❏	❏	❏
If the United States/ France does not conduct the drone strike, the terrorist attack will **result in many casualties**?	❏	❏	❏	❏	❏
If the United States/ France does not conduct the drone strike, will the terrorist group be **emboldened to plan more attacks**?	❏	❏	❏	❏	❏
If the United States/ France does not conduct the drone strike, will the security of other states **suffer**?	❏	❏	❏	❏	❏
If the United States/ France does not conduct the drone strike, will its credibility, prestige, or reputation **suffer**?	❏	❏	❏	❏	❏

a. In your estimate, if the United States/France <u>does not</u> conduct the drone strike, what are the chances that each of the following things will happen?

Here is the situation again for your reference:

- A terrorist group has enjoyed sanctuary in a fragile country in a volatile region of the world.
- The terrorist group has planned to attack a western country with the intent of killing civilians.
- Given the intelligence, the United Nations has/has not authorized the use of force, including armed drones, to prevent the attack.
- The United States/France decides to use a drone strike to kill the terrorists responsible for planning the attack. Although the strike removes the terrorists, the terror group's leaders claim the operation also resulted in collateral damage, including civilian casualties. American/French officials adamantly deny this claim.

b. In your estimate, if the United States/France <u>does</u> conduct the drone strike, what are the chances that each of the following things will happen?

To review, here is the situation again:

- A terrorist group has enjoyed sanctuary in a fragile country in a volatile region of the world.
- The terrorist group has planned to attack a western country with the intent of killing civilians.
- Given the intelligence, the United Nations has/has not authorized the use of force, including armed drones, to prevent the attack.
- The United States/France decides to use a drone strike to kill the terrorists responsible for planning the attack. Although the strike removes the terrorists, the terror group's leaders claim the operation also resulted in collateral damage, including civilian casualties. American/French officials adamantly deny this claim.

c. To what degree do you believe the drone strike was compatible with international law?

  - Very Compatible
  - Somewhat Compatible
  - Neither Compatible Nor Incompatible
  - Somewhat Incompatible
  - Very Incompatible

	Almost no chance	25% chance	50–50 chance	75% chance	Nearly 100% certain
If the United States/ France does conduct the drone strike, will the terrorist attack be **prevented**?	❏	❏	❏	❏	❏
If the United States/ France does conduct the drone strike, will the terrorist group be **deterred from conducting future attacks**?	❏	❏	❏	❏	❏
If the United States/ France does conduct the drone strike, will the security of other states be **enhanced**?	❏	❏	❏	❏	❏
If the United States/ France does conduct the drone strike, will it **minimize harm to its own soldiers**?	❏	❏	❏	❏	❏
If the United States/ France does conduct the drone strike, will it **minimize the financial costs of using force abroad**?	❏	❏	❏	❏	❏
If the United States/ France does conduct the drone strike, will its relations with other countries **suffer**?	❏	❏	❏	❏	❏

Again, for your reference:

- A terrorist group has enjoyed sanctuary in a fragile country in a volatile region of the world.
- The terrorist group has planned to attack a western country with the intent of killing civilians.
- Given the intelligence, the United Nations has/has not authorized the use of force, including armed drones, to prevent the attack.

- The United States/France decides to use a drone strike to kill the terrorists responsible for planning the attack. Although the strike removes the terrorists, the terror group's leaders claim the operation also resulted in collateral damage, including civilian casualties. American/French officials adamantly deny this claim.

d. To what extent do you agree that there was a moral obligation to intervene?

- Strongly Agree
- Somewhat Agree
- Neither Agree Nor Disagree
- Somewhat Disagree
- Strongly Disagree

Here is the situation one last time:

- A terrorist group has enjoyed sanctuary in a fragile country in a volatile region of the world.
- The terrorist group has planned to attack a western country with the intent of killing civilians.
- Given the intelligence, the United Nations has/has not authorized the use of force, including armed drones, to prevent the attack.
- The United States/France decides to use a drone strike to kill the terrorists responsible for planning the attack. Although the strike removes the terrorists, the terror group's leaders claim the operation also resulted in collateral damage, including civilian casualties. American/French officials adamantly deny this claim.

e. In your estimate, how likely is it that other states would have helped carry out the drone strike?

- Very Likely
- Somewhat Likely
- Neither Likely nor Unlikely (or 50/50 chance)
- Somewhat Unlikely
- Very Unlikely

Follow-up Demographic and Dispositional (Conservatism, Ethnocentrism, Foreign Policy Orientation, Internationalism, Religiosity) Variables

Programming Note: The following questions (7a–7o) are randomized.

a. What is your annual income?

- Less than $10,000
- $10,000 to $24,999

- $25,000 to $49,999
- $50,000 to $74,999
- $75,000 to $99,999
- $100,000 or more

(The following questions are designed to assess respondents' conservatism.)

b. Generally speaking, do you usually think of yourself as a . . .?

- Democratic
- Independent
- Republican
- Other
- I am not sure

c. Generally speaking, do you think of yourself as . . .?

- Extremely Liberal
- Liberal
- Slightly Liberal
- Moderate, middle of the road
- Slightly Conservative
- Conservative
- Extremely Conservative
- I am not sure

d. Did you vote in the 2020 U.S. presidential election?

- Yes
- No

Programming Note. Those respondents who answer "Yes" to question 7d will receive the following question:

e. Which candidate did you vote for?

- Joseph Biden
- Donald Trump
- Other

(The following question is designed to assess respondents' ethnoc-entrism.)

f. To what extent do you agree with the following statement? "When jobs are scarce, employers should give priority to the people of this country over immigrants."

- Strongly Agree
- Agree
- Neither Agree Nor Disagree

- Disagree
- Strongly Disagree

(The following questions are designed to assess respondents' foreign policy orientation or preference for the use of force abroad.)

g. To what extent do you agree with the following statement? "The use of military force has a role to play in international affairs."

- Strongly Agree
- Agree
- Neither Agree Nor Disagree
- Disagree
- Strongly Disagree

h. To what extent do you agree with the following statement? "The use of military force only makes problems worse."

- Strongly Agree
- Agree
- Neither Agree Nor Disagree
- Disagree
- Strongly Disagree

(The following questions are designed to assess respondents' internationalism or awareness of global politics.)

i. To what extent do you agree with the following statement? "Great powers such as the U.S. and China need to play an active role in solving problems around the world."

- Strongly Agree
- Agree
- Neither Agree Nor Disagree
- Disagree (2)
- Strongly Disagree

j. To what extent do you agree with the following statement? "International organizations are taking away too much power from countries."

- Strongly Agree
- Agree
- Neither Agree Nor Disagree
- Disagree
- Strongly Disagree

(The following questions are designed to assess respondents' religiosity.)

e. How would you describe your religious affiliation today? (Please select only one.)

- Protestant Christian
- Catholic
- Other Christian
- Jewish
- Muslim
- Buddhist
- Hindu
- Atheist
- No formal religious affiliation
- Other

Programming Note. Those respondents answering "Catholic," "Other Christian," or "Protestant Christian" to question 7k will receive the following question:

f. Are you a "born-again" or evangelical Christian?

- Yes
- No

g. How important is religion in your life?

- Very Important
- Somewhat Important
- Neutral
- Somewhat Unimportant
- Very Unimportant

(The following questions are designed to assess respondents' military status.)

h. Have you ever served or are you currently serving in the US military (including any component—Active Duty, National Guard, Reserves)?

- Yes
- No

i. Has another member of your immediate family (other than you) ever served in the US military (including any component—Active Duty, National Guard, Reserves)?

- Yes
- No

Table C.1 Summary Statistics

		France (N = 909)		US (N = 914)		Full Sample (N = 1823)	
		Number	*Percent*	*Number*	*Percent*	*Number*	*Percent*
Gender	Female	472	51.9	474	51.9	946	51.9
	Male	435	47.9	437	47.8	872	47.8
	Other	2	0.2	3	0.3	5	0.3
Age	18–24	116	12.8	98	10.7	214	11.7
	25–34	123	13.5	141	15.4	264	14.5
	35–44	167	18.4	207	22.6	374	20.5
	45–54	142	15.6	106	11.6	248	13.6
	55–65	251	27.6	168	18.4	419	23.0
	Over 65	110	12.1	194	21.2	304	16.7
Education	Less than high school	136	15.0	87	9.5	223	12.2
	High school (or equivalent)	374	41.1	256	28.0	630	34.6
	Some college	132	14.5	161	17.6	293	16.1
	2-year college degree	60	6.6	93	10.2	153	8.4
	4-year college degree	23	2.5	203	22.2	226	12.4
	Professional degree	184	20.2	114	12.5	298	16.3
Strike Attributes	Control	179	19.7	183	20.0	362	19.9
	Other country, multilateral	185	20.4	181	19.8	366	20.1
	Other country, unilateral	183	20.1	182	19.9	365	20.0
	Own country, multilateral	180	19.8	188	20.6	368	20.2
	Own country, unilateral	182	20.0	180	19.7	362	19.9

Note: This table shows descriptive statistics for our sample. We dropped respondents who did not identify as men or women for ease of analysis. The statistical power for this category would have been too low for any meaningful analysis.

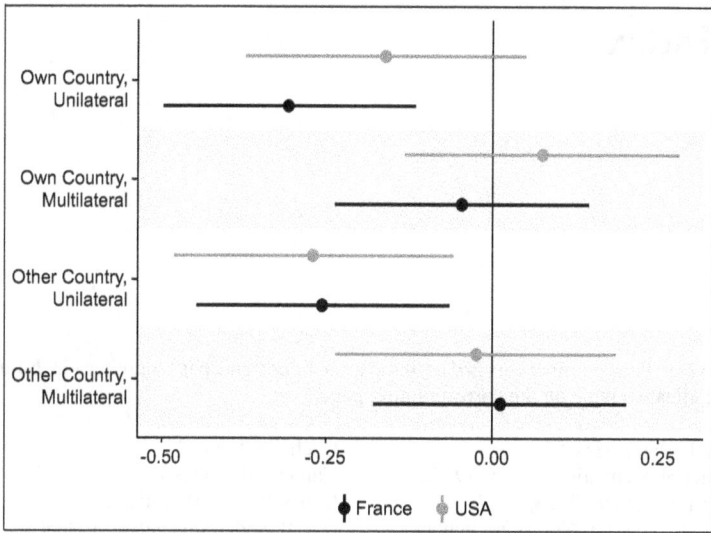

Figure C.2 Impact of Strike Attributes on Public Perceptions of Legitimacy by Country

Note: This figure shows how different strike attributes, in this case the country conducting the strike as well as international approval, shape public perceptions of legitimate drone warfare in a cross-national setting.

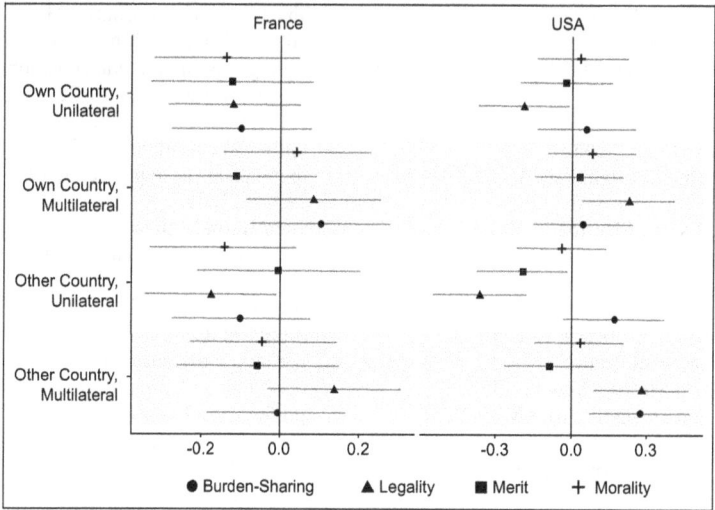

Figure C.3 Impact of Strike Attributes on Mechanisms

Note: This figure shows how microfoundations work to shape public perceptions of legitimate drone warfare in a cross-national setting.

Index